The Haynes Fuel Injection Manual

by Don Pfeil and John H Haynes
Member of the Guild of Motoring Writers

The Haynes Workshop Manual
for automotive fuel injection systems

(10T1 – 482)

Haynes Publishing Group
Sparkford Nr Yeovil
Somerset BA22 7JJ England

Haynes North America, Inc.
861 Lawrence Drive
Newbury Park
California 91320 USA

Acknowledgements
We are grateful for the help and cooperation of AB Volvo, American Motors Corporation, Bayerische Motoren Werke (BMW), Chrysler Corporation, Ford Motor Company, General Motors Corporation, Mercedes-Benz, Nissan Motor Company, LTD. and VW/Audi for their assistance with technical information, certain illustrations and photos.

© **Haynes North America, Inc. 1986**
With permission from J. H. Haynes & Co. Ltd.

A book in the **Haynes Automotive Repair Manual Series**

Printed in the USA

All rights reserved. No part of this book may be reproduced or transmitted in any form or by any means, electronic or mechanical, including photocopying, recording or by any information storage or retrieval system, without permission in writing from the copyright holder.

ISBN 0 85696 482 4

Library of Congress Catalog Card Number 86-82626

While every care is taken to ensure that the information in this manual is correct, no liability can be accepted by the authors or publishers for loss, damage or injury caused by any errors in, or omissions from, the information given.

Contents

Chapter 1
Introduction 5

Chapter 2
Applications 23

Chapter 3
Bosch Fuel Injection Systems 29

Chapter 4
General Motors Fuel Injection Systems 62

Chapter 5
Ford Fuel Injection Systems 90

Chapter 6
Chrysler Fuel Injection Systems 115

Index 139

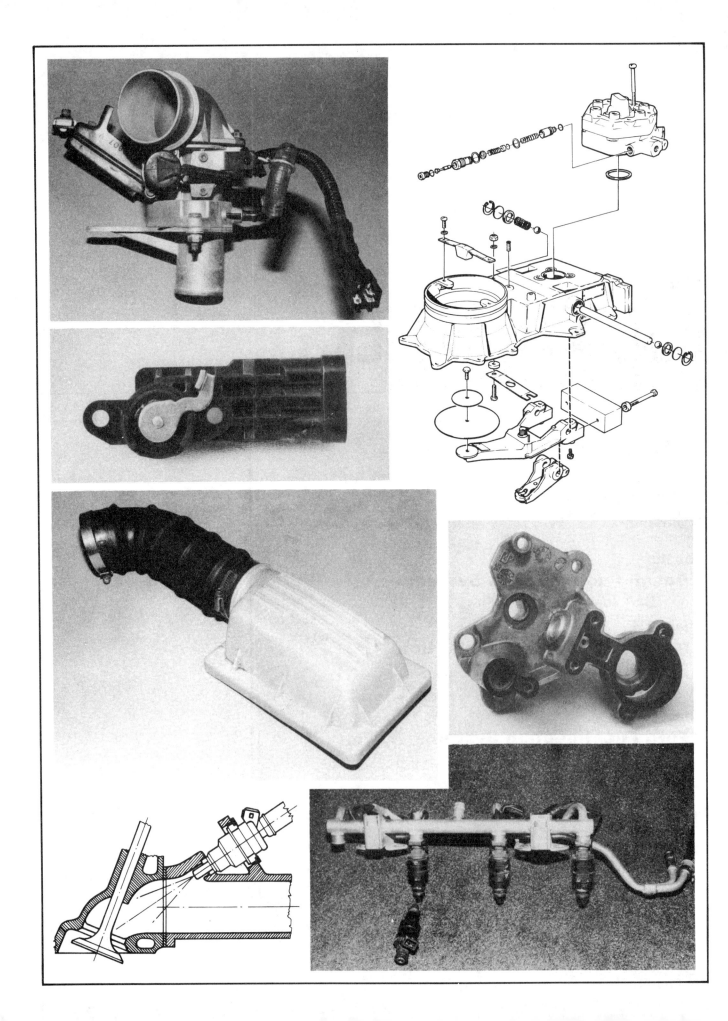

Introduction

Introduction 1
   Why fuel injection?
   What is it?
   Types of fuel injection

Description 2
   Mechanically timed injection
   Electronically timed injection
   Continuous flow injection
   Throttle body injection

Information (sensor) components 3
   The computer
   Air flow sensor
   Air mass sensor
   Throttle position sensor
   Oxygen sensor
   Coolant temperature sensor
   Fuel temperature sensor
   Detonation sensor
   Speed sensor
   Boost sensor
   Manifold and outside barometric pressure sensors
   Intake charge temperature sensor
   Crankshaft position sensor

Operating components 4
   Fuel pressure regulator
   Electronic control unit
   Air control valve
   Fuel injectors
   Cold start valve
   Thermo-time switch
   Warm-up regulator
   Auxiliary air regulator
   Fuel distributor
   Fuel accumulator
   Idle speed bypass valve

1 Introduction

Why fuel injection?

Fuel injection is one of those modern, high-tech ideas that has become almost a "required option" if high performance is desired in a car. But, in fact, fuel injection has been around for decades, and has been the standard method of delivering fuel into both racing and aviation engines since before the Second World War. In fact, the first airplane, the Wright Flyer that flew at Kitty Hawk is 1903, had an engine equipped with fuel injection.

Why, then, is fuel injection just now coming into widespread use on production automobiles? Basically, for two reasons. The first is the development of the inexpensive computer, without which fuel injection, and many of the related emissions-control devices on a modern car, would be either impossible to utilize or so expensive it couldn't be used on a high production automobile.

The second reason fuel injection has come into use also relates to emissions. Each year the amount of pollutants the automobile is allowed to emit is reduced, and after several years of developing add-on devices that either reduce pollutant production or neutralize pollutants, manufacturer's turned to attempts to make the engine run more efficiently, thereby producing less noxious emissions. And the key to efficiency is a well-controlled fuel/air mixture which can be maintained at close to ideal levels over a broad range of engine operating conditions and speeds.

What is it?

Fuel injection is simply a method of delivering a mixture of fuel and air to the engine's cylinders. This is the same thing that is done by the carburetor, but with a fuel injection system it can be done much more efficiently.

To burn properly in an engine, gasoline has to be mixed with air in a ratio of between 12 to 1 and 16 to 1. Unfortunately, a spread of ratios like that can burn too much fuel if the mixture is too rich, destroy valves and even pistons if it is too lean, cost power if the mixture is too far on either side of the "best" ratio, and create clouds of pollution. These days, to get maximum horsepower out of the smaller engines being built, to reduce emissions to an absolute minimum, to get the best mile-per-gallon figures possible and to keep the engine running for a reasonable length of time between tune ups and overhauls, a ratio of very near 14.7 to 1 is required — at all speed ranges and throttle openings. And this is where the carburetor proves inferior to fuel injection.

A carburetor, even the best one around, is a basically very simple device, surrounded by a large number of add-on systems to correct the deficiencies that are a part of the basic carburetor design — deficiencies which cannot be designed out simply because they are inherent in the carburetor method of mixing fuel and air.

Any carburetor has to be several systems built into one body. There has to be a system to provide fuel and air to the engine in the proper proportions for idling. This system would also need some method of enriching the mixture for cold starting and running until the engine warms up. Another system is needed to provide fuel in the acceleration mode, when, to avoid stumble when the throttle is suddenly opened, a richer mixture is required. Then there is still another system required to provide fuel and air properly mixed for steady cruising speeds.

Designing the systems for each of these modes of operation requires a complex device, but it can be (and is) done. Each system works exactly as required. Unfortunately, no engine goes instantly from idle to acceleration, or from acceleration to cruising, nor does any engine stay at one set speed.

Fuel Injection

If it did, there would be no need for complicated carburetors or fuel injection systems. And it is when operating modes overlap, where two or more systems are trying to work at the same time to give the engine a proper fuel/air mixture ratio, that the operation of the carburetor becomes inefficient.

Modern fuel injection is the answer to the carburetor's problems. Through sensors mounted on various parts of the engine, the engine computer can determine just what fuel/air mixture is required at any given moment. The air intake system (throttle valve housing) is simply an air gate, rather than a mixing device, so the incoming air charge can be measured accurately and there is no concern about introducing a uniform mixture of gasoline into that air flow.

The fuel is injected into the air mass just outside the intake valve, so atomization can be controlled precisely, delivery can be timed, and there is no problem with the cylinders furthest from the carburetor being too lean or the ones closest being too rich. And the computer can "read" the temperature of the engine, the rpm, the mass of air flowing through the intake system, how far the throttle is opened, the air pressure inside the intake manifold and the outside air pressure, and many other factors, adjusting the fuel mixture to meet the requirements of the engine, over a thousand times per second!

Types of fuel injection

There are three basic types of fuel injection used on modern automobiles with gasoline engines. They are:
- Timed fuel injection
- Continuous fuel injection
- Throttle body fuel injection

In addition, there are further subdivisions within each of these three categories.

The Timed Fuel Injection system may be either mechanical or electronically timed.

Fuel injection is a more efficient method of delivering fuel to the cylinder because the fuel is injected directly into the intake manifold, rather than being mixed with air at the carburetor then routed through the intake manifold runners before entering the cylinder

The Timed Fuel Injection system may be either mechanical or electronically timed.

The Throttle Body Injection system is almost always electronically timed, but there are some systems which utilize a continuous spray of fuel from a spray bar, rather than an electrically operated solenoid type injector, controlling the fuel flow by electronic regulation of the fuel pump.

1. FUEL INJECTOR
2. INTAKE MANIFOLD
3. INTAKE VALVE
4. ELECTRICAL TERMINAL
5. "O" RING
6. FUEL RAIL

2 Description

Mechanically timed injection

These days most mechanically timed injection systems are used on diesel engines, but there are still some units in use on gasoline engines. This form of injection was, at one time, extremely popular on race car engines, since it was able to deliver identical amounts of fuel, in large quantities, directly to each cylinder just as the intake valve was opening.

Components and Operation

There are two basic types of mechanically timed systems — the high pressure system with a metering unit and the low pressure type with an injection pump.

With the first type the fuel is delivered from the tank to the injection metering pump by an electric pump at very high pressure, often in the 100 to 125 psi range. A relief valve in the metering system bleeds off fuel not required by the injectors and returns it to the tank, so the pressure remains constant at all engine speeds.

Inside the metering unit a rotor, driven directly by the engine, distributes fuel to the individual injectors at the right time. The rotor is driven at one-half crankshaft speed, so the opening in the rotor delivers the fuel to any given injector only on the intake stroke, and not on the exhaust stroke. The injector nozzles are spring-loaded, remaining closed until forced open by the high pressure fuel delivery and shutting again as soon as the rotor cuts off the fuel pressure to that injector.

The second type of mechanical metering system, and the one which is often used on diesel engines, delivers the fuel from the tank at low pressure to a fuel injection pump, which is actually a row of pistons and cylinders, each piston and cylinder delivering a pulse of fuel to the cylinder at the precise moment for maximum efficiency. In effect, the injection pump is a small duplicate of the engine, but it pumps straight fuel rather than an air/fuel mixture.

Electronically timed injection

As with mechanically timed injection, the fuel is delivered to an injector, usually just above the intake valve, by a high pressure pump. But, instead of the injector being forced open by the pressure of the fuel "shot" from the pump, the injector is electronically controlled and electrically opened, usually at the command of a computerized control system which responds to signals from a variety of engine sensors.

This small computer reads the operating conditions, such as the throttle position, the amount of air entering the intake, the temperature of the engine, the rpm, and a variety of other parameters, then it sends a signal to the injector, which is actually a small electrical solenoid with a small, precision machined valve inside.

The signal opens the solenoid valve, which sprays the pressurized fuel into the intake port. Since the pressure is held constant by the fuel pump and pressure metering system, precise control of the amount of time the injector is held open by the computer control unit regulates the amount of fuel delivered.

Some modern electronically controlled fuel injection systems have become so sophisticated in the design of their computer controls that they can sense a detonation condition in a cylinder, and vary both the ignition timing and the fuel mixture for that individual cylinder to eliminate the detonation problem!

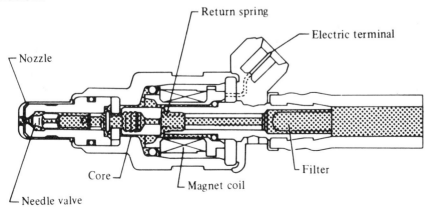

Electronically timed fuel injection uses a solenoid-type injector, where an electrical signal opens the injector valve, allowing the fuel, which is under constant pressure, to spray into the intake manifold

Fuel Injection

Continuous flow injection

Continuous flow fuel injection is much simpler in design than the timed systems, and therefore is somewhat less precise and efficient, but the lack of complexity also makes it much easier to maintain and less expensive to produce, which has been one of the key factors in the increase in low to medium price vehicles equipped with fuel injection.

As indicated by the name, the injection of fuel into the intake system is continuous, rather than in the individual bursts of the timed system. Once the engine is started, the injectors remain open. Fuel is supplied to a metering unit, and a valve in that unit controls the amount of fuel delivered to the injectors. A sensor in the intake system measures the mass of air entering the engine, and this measurement is what controls the amount of fuel delivered.

The individual injectors are located near the intake valves, just as with the timed systems, but there is a spray of fuel even when the valve is closed, which slightly reduces the efficiency of the system.

In a continuous flow system, an air flow sensor (A) monitors the amount of air entering the engine, and the fuel distributor (B) delivers the required amount of fuel to the cylinders

Throttle body injection

Throttle body injection, used extensively on U.S. produced cars, is a hybrid system, utilizing the central fuel mixing feature of a carburetor, and the electronically controlled fuel valving of a timed injection system.

As with the electronically timed systems, throttle body injection uses a solenoid-type injector, controlled by a computer system which reads the

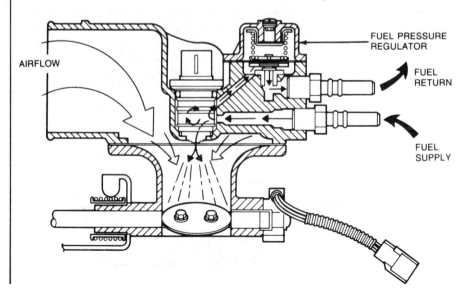

A throttle body system uses an electrically-controlled solenoid, essentially the same as the unit used in a electronically timed injection system, but instead of having one injector per cylinder, the TBI system uses one centrally mounted injector in a carburetor-like throttle body

Components and Operation

engine operating parameters and regulates the open time of the injector to control the mixture. However, instead of having an injector positioned at each port, the throttle body systems utilizes one or two injectors inside what amounts to a carburetor body, spraying the fuel into an essentially conventional intake manifold.

The advantage of the throttle body injection system over a conventional carburetor is the elimination of a float system, idle, acceleration (power) and main metering systems, the choke assembly and their replacement with accurate fuel metering through the solenoids.

3 Information (sensor) components

The computer

Without the computer, usually referred to as a microprocessor, there probably wouldn't be any fuel injection on today's engines — or at least not on production engines. Fuel injection unregulated by computer would be both too expensive and too limited in capability (more suited for racing applications) for moderately priced vehicles.

If you didn't already know what it looked like, someone has undoubtedly pointed out a box under your hood and told you that it was the computer that controls most of the functions of your engine. The box is probably six or so inches on a side and an inch thick, and, in general, it is the computer.

Actually, the computer is a very small component inside that box, called a "chip." Often less than a quarter-inch square and a small fraction of an inch thick, the chip usually contains thousands of separate circuits, and it is those circuits which control the operation of your fuel injection system, as well as the emissions systems and ignition system.

While the computer (the chip) has those thousands of circuits, and is capable of lightening-fast decision making, it can't actually do anything, and this includes finding out what is going on around it. For that, the computer needs a wide array of sensors to feed information to the chip. Not all of these sensors are used with every fuel injection system, but all injection systems use several sensors. These sensors include:

The computer, usually referred to as the Electronic Control Module (ECM) or Electronic Control Unit (ECU), is the heart of the electronic and throttle body injection systems. The computer 'reads' operating data from sensors and adjusts the amount of fuel delivered to the cylinders for optimum performance

Fuel Injection

Air flow sensor

The air flow sensor is a precisely counterbalanced plate located in the air venturi. As the throttle is opened, air flowing through the venturi lifts the air flow sensor plate. The more air there is flowing into the engine, the higher the plate is lifted.

The plate, in turn, is connected to the control plunger in the fuel distributor, which regulates the amount of fuel being delivered to the injectors.

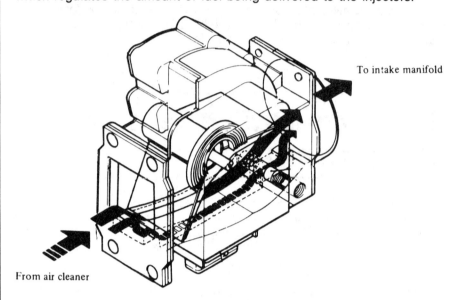

To intake manifold

From air cleaner

Various types of air flow measuring devices are used to determine the amount of fuel needed by the engine, but most utilize a movable plate or vane which is displaced by the air being drawn into the engine

Air mass sensor

Installed in the air intake passage of some fuel injection systems is a platinum wire or a thin film which is heated by a current from an electronic amplifier, which measures the resistance to the current flow. The hotter the wire or film becomes, the greater the resistance. The amplifier controls the current flow so as to maintain the wire at an even temperature, and therefore at an even level of resistance.

Air flowing past the wire cools it, and the more air flowing past the wire, the greater the cooling effect. As the wire cools, the amplifier has to send more current through the wire to keep it at the set temperature, and this greater current flow is signalled to the ECU, which injects more fuel to match the greater air flow.

At the same time the signal is being sent by the air mass sensor, other signals are coming into the electronic control unit from the throttle position sensor, the engine rpm sensor and the air temperature sender. All of these signals are integrated by the ECU to determine how long the injectors should be held open, and on some systems the signals are also used to adjust the ignition timing (advance) for optimum engine operation.

Throttle position sensor

The throttle position sensor, or throttle switch, is mounted on the throttle chamber, moves in accordance with the movement of the accelerator pedal, and sends a position signal to the ECU. In most applications the sensor monitors idle and wide-open throttle positions, although some have a third position, signalling mid-range throttle opening.

When the throttle is in the idle position, the throttle sensor signals the ECU that idle enrichment is needed, and, when the ignition switch is shut off, the

Components and Operation

throttle sensor sends the signal which shuts off the fuel flow from the tank by cutting the electrical power to the fuel pump. The sensor also cancels the imput to the oxygen sensor during coast-down to keep the system from being affected by the excessive oxygen in the system at that time.

The throttle position sensor is mounted on the throttle body or throttle chamber, where it monitors both idle and wide-open throttle conditions, sending signals to the Electronic Control Unit to modify the fuel flow as necessary

The oxygen sensor, located in the exhaust stream, enables the ECU to control the mixture, and therefore the level of emissions.

The oxygen sensor constantly monitors the amount of unburned oxygen in the exhaust, and the signal sent to the ECU is matched against the ideal. A signal from the ECU richens or leans the mixture as necessary for proper emissions control.

The oxygen sensor only operates after it has been heated by the exhaust stream, so, to provide more rapid actuation of the system and thereby reduce the time when emissions are essentially uncontrolled, some oxygen sensors contain an electrical heating element. This type of oxygen sensor can be identified by the use of three wires, rather than the two used on unheated sensors. The two types of sensors cannot be interchanged.

Oxygen sensor

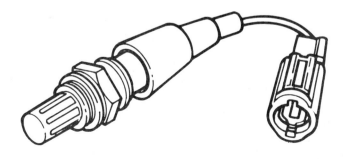

The oxygen sensor is mounted in the exhaust stream, where it measures the amount of oxygen in the exhaust gas. This measurement is sent to the Electronic Control Unit, where the information is used to determine the amount of fuel needed to maintain an optimum fuel/air mixture

Fuel Injection

Coolant temperature sensor

When the engine is cold, additional fuel is required, and, just as the choke on a carburetor enrichens the mixture by reducing the amount of air taken into the engine, the coolant temperature sensor signals the ECU to hold the injectors open longer when the engine is cold, passing more fuel into the engine.

Two methods are used to signal the engine temperature to the ECU. On many engines a temperature sensor is inserted into the cooling system, much like the coolant temperature probe which activates the temperature gauge or warning light. As the coolant warms the extra enrichment is reduced, until the engine is at normal operating temperature.

On some engines, especially later models and those with turbochargers, a cylinder head temperature sensor is used instead of the coolant sensor. In this case the actual temperature of the engine is signalled to the ECU, which gives a somewhat more accurate indication of operating conditions than the coolant temperature sensor.

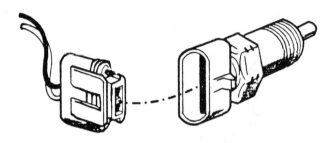

A richer mixture is required when the engine is cold, and the coolant temperature sensor is utilized by the Electronic Control Unit to determine when this richer mixture is necessary

Fuel temperature sensor

As gasoline heats, it expands, and therefore warmer fuel would result in a leaner mixture. Because of this some systems incorporate a fuel temperature sensor, generally built into the fuel pressure regulator, which signals to ECU to enrich the mixture whenever the fuel temperature exceeds a certain value.

Detonation sensor

On some engines, generally those installed in high performance models and those equipped with a turbocharger, a detonation sensor is included in the fuel injection sensor system to detect engine knock, caused be preignition of fuel with too low an octane rating or preignition caused by excessive temperature in the cylinder, usually caused by excessively advanced ignition timing.

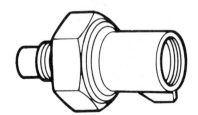

When detonation is detected by the vibration sensor, which is usually installed in the cylinder block, a signal is sent to the ECU to retard the ignition timing, richen the mixture, or both.

The detonation sensor detects pre-ignition conditions inside the cylinder and signals the Electronic Control Unit to richen the mixture and/or retard the ignition timing until the abnormal condition is corrected

Components and Operation

Speed sensor

A road speed sensor is included with some systems to give the electronic control unit a modifying signal for the engine rpm sensor. The sensor is mounted in the speedometer unit, and is made up of a reed-type switch on indicator needle type speedometers or a LED, photo diode and shutter on digital type speedometers.

Boost sensor

On turbocharged models a boost sensor is incorporated in the system. The sensor signals the ECU under high boost conditions to enrichen the mixture. On some models the ECU also controls the wastegate to limit boost and cuts off the fuel delivery to the injectors to prevent engine damage.

Manifold and outside barometric pressure sensors

The outside barometric pressure sensor signals changes in pressure due to temperature or altitude change, which would alter the amount of air flowing into the engine and therefore the amount of fuel required. The manifold pressure sensor registers changes resulting from changes in engine load, engine speed and throttle opening, sending those signals to the electronic control assembly. The ECU combines the two pressure signals to determine the amount of fuel needed under the current operating conditions.

Intake charge temperature sensor

As the temperature of the incoming air changes, the density of that air also changes. In addition, the spray of fuel into the incoming air from throttle body mounted injectors will also change the temperature of the fuel charge, increasing charge density. Signals recording these changes in density are sent to the electronic control assembly by the intake charge temperature sensor to modulate the amount of fuel injected, maintaining the proper fuel/air mixture for optimum engine performance and minimum emissions.

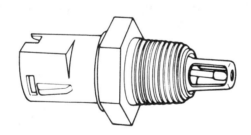

The intake temperature sensor monitors the temperature of the air being drawn into the engine and the Electronic Control Unit uses the signal from the sensor to increase or decrease the amount of fuel delivered to adjust for changes in the density of the incoming air

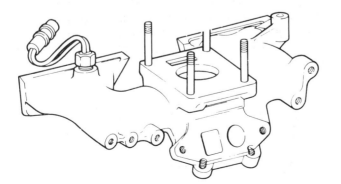

Some intake temperature sensors are mounted in the intake manifold, while others are installed in the air cleaner assembly

Fuel Injection

Crankshaft position sensor

A magnetic pickup on the nose of the crankshaft or a rotating plate in the distributor housing provides the electronic control assembly with a crankshaft position signal which is used both as an ignition timing reference and to control injector operation.

4 Operating components

Once the sensors have sent information to the computer, the actual devices which make up the fuel injection system go to work, with their operation controlled by signals from the computer. Again, not all fuel injection systems use all of the components listed here, but your system includes some of these components.

Fuel pressure regulator

A key component of the system is the fuel pressure regulator, which works with the fuel pump to maintain a steady pressure relationship between the fuel line side of the injectors and the intake manifold.

Since on most systems the fuel is metered by the ECU by altering the amount of time the injector is held open, the volume of fuel injected would vary if there was a pressure difference between the fuel supply and the intake manifold under different conditions. When there is a high vacuum in the intake manifold, such as at high rpm, the pressure in the fuel system must be reduced. When there is low pressure in the intake manifold, such as at low speed, full throttle operation, the pressure in the fuel line must be increased. On the older type systems where the fuel is metered by the actual amount of fuel delivered to the injector by a injection pump, the pressure is equally critical since the system has to operate under much higher pressure than a solenoid injector type system.

On solenoid systems, the essential element is that the pressure differential remain the same under all circumstances, so that only the time the injector is open determines the amount of fuel injected. In order to maintain the desired fuel pressure, excess fuel is returned to the fuel tank by a separate line from the fuel pressure regulator.

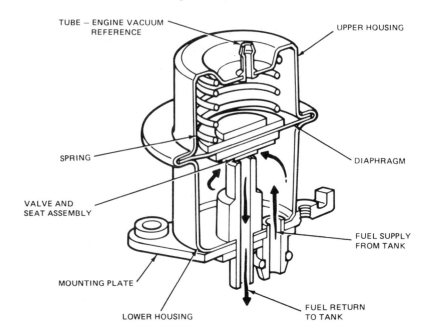

The fuel pressure regulator maintains a constant pressure differential between the fuel line and the intake manifold pressure so that only the time the injector is open controls the amount of fuel delivered to the engine

Components and Operation

Electronic control unit

The Electronic Control Unit (ECU) receives the signals from the sensors associated with the fuel injection unit, such as the throttle position sensor, air mass sensor, etc., and sends out the signals to the injectors to control the amount of fuel injected into the cylinders.

Integrated into the ECU is the Programmable Read Only Memory unit (PROM), which contains the instructions for that particular vehicle and fuel injection design. Each manufacturer has different operating parameters for the fuel injection system, and therefore each has different PROMS. Because of this, the ECU cannot be interchanged between different makes of vehicles, even when those makes all use an essentially similar system and in many cases the ECU cannot even be interchanged amoung different models of the same manufacturer.

The ECU is a solid state device (no moving parts), and it is very seldom that an ECU will fail in service. This is not to say, however, that an ECU cannot go "bad." But when an ECU does fail, it is almost invariably because something outside the unit has failed, interferring with the signals into or out of the ECU. In this case, it is common for the ECU to self-destruct, and since it is a very expensive piece of equipment, it is advisable to exercise extreme caution when working with any components connected to the ECM.

The two key precautions to take when working with the ECU or ECU-connected components are never to disconnect the ECU with the ignition switch in the On position and always take extreme care with connectors to make sure that connecting pins enter only the correct receptacles.

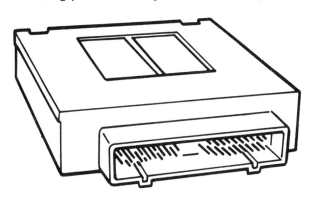

The electronic control unit receives signals from various engine sensors, interprets the data to determine the amount of fuel needed by the engine, and signals the injectors to remain open just long enough for that amount of fuel to be injected

Air control valve

An air control valve is used to bypass the throttle valve, setting the idle speed. The amount of air sent through the air control valve is regulated by the ECU.

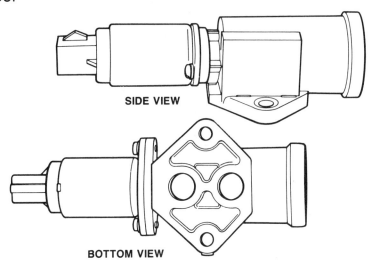

The air control valve bypasses the throttle valve to control the idle speed. The amount of air passed by the air control valve is controlled by the Electronic Control Unit

Fuel Injection

Fuel injectors

On most systems the fuel injector is an electrical solenoid, controlled by the ECU. Fuel under pressure is supplied to the injectors, and a pressure differential is maintained between the fuel line pressure on one side of the injector and the manifold pressure on the other side.

Inside the injector unit is a coil, and when current is supplied to the coil the injector valve opens, allowing fuel to pass through the unit and into the intake manifold. The ECU controls the current reaching the injector, and the longer the ECU sends an "open" signal, the more fuel is injected into the engine.

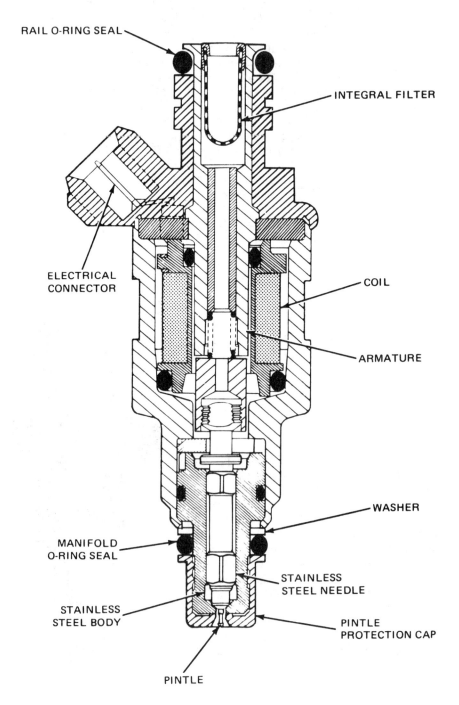

The most commonly used injector is the electrically-operated solenoid, where a signal from the Electronic Control Unit opens a valve, allowing the fuel, which is maintained at a set pressure by the regulator, to spray into the intake manifold

Components and Operation

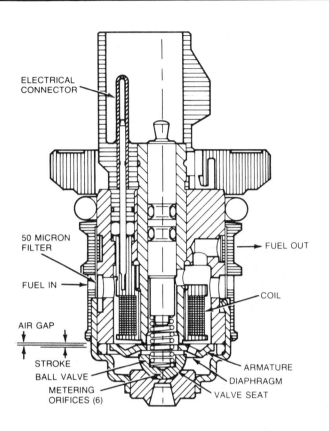

This design of injector looks completely different than the common electrically-operated unit, but the operation is basically the same, with an electrical signal, sent by the Electronic Control Unit, controlling the amount of time the injector is held open

The fuel injectors for some systems are mechanical units, spring loaded and set to open at anything over a preset pressure, usually approximately 50 psi. The injectors spray fuel into the intake manifold upstream from the intake valves in a constant stream, so long as the fuel pressure remains high enough to hold the injector open.

The injectors have no metering function. They are either open or closed, and the amount of fuel delivered by the injector is determined by the pressure allocated by the fuel distributor. The higher the pressure, the more fuel.

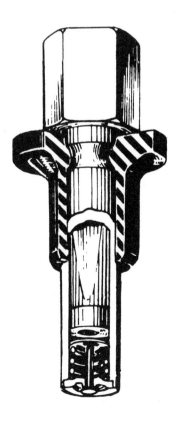

In a continuous flow system the injectors are spring-loaded and are opened by fuel pressure, spraying a constant stream of fuel into the intake manifold. The amount of fuel injected into the engine with this system is controlled by the fuel distributor

Fuel Injection

Cold start valve

Under cold starting conditions extra fuel is required by the engine, and it is the function of the cold start valve to supply that fuel. The valve is actually a fuel injector, activated by a solenoid, and the fuel is injected at a central location into the intake manifold to supplement the fuel being delivered by the individual injectors.

An engine temperature sensor and an electrically heated bimetallic strip are used to control the cold start valve, which is automatically shut off after a limited amount of operation to prevent flooding.

On many injection systems a cold start valve is used to supply extra fuel to the engine for faster warm-up and even running while the engine is cold

The cold start valve is not used to supply the total fuel supply to the engine under cold operating conditions, but only to supplement the fuel being delivered by the fuel injectors

Thermo-time switch

The thermo-time switch regulates the amount of time the cold start valve is energized and prevents the opening of the cold start valve and subsequent mixture enrichment when the engine is warm.

The "thermo" section of the switch is a heater coil. When the starter is energized, current is passed through the thermo-time switch and into the cold start valve. As the current passes through the switch, the heater coil warms a bimetallic switch to the point where it opens the circuit, eliminating the extra fuel flow enrichment through the cold start valve.

Components and Operation

The length of time the circuit is energized is controlled by the engine coolant temperature. The lower the coolant temperature, the longer it will take the heater coil to warm the bimetallic switch.

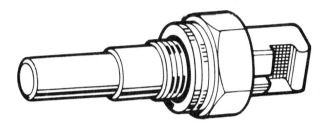

The thermo-time switch controls the amount of time the cold start valve is held open by measuring the coolant temperature and utilizing a time function to prevent an over-rich condition

Warm-up regulator

The warm-up regulator also acts to enrich the mixture during cold engine operation by causing the fuel distributor to open the fuel metering ports an extra amount, delivering more fuel to the fuel injectors.

An electrically heated bimetallic spring works against the control pressure circuit diaphragm, increasing the fuel pressure in the fuel distributor by returning less excess fuel to the tank. As the bimetallic spring is heated, the pressure of the spring against the control pressure diaphragm is lessened, until the engine is warmed and the warm-up regulator is removed from the circuit.

Auxiliary air regulator

The auxiliary air regulator acts as an air control system during cold engine operation and start-up. When the engine is cold, it bypasses extra air around the the throttle plate, which is read as extra air flow by the ECU, which signals for additional fuel.

A bimetallic spring activated by coolant temperature and an electrical heater coil to prevent excessive operation of the auxiliary air regulator control a rotating disc inside the throttle chamber. The disc has a hole in it, and when the hole lines up with a matching hole in the throttle chamber bypass, extra air is inducted into the engine. As the bimetallic spring heats up, it rotates the disc, gradually closing off the hole until all the air is passing into the engine through the main throttle chamber passage.

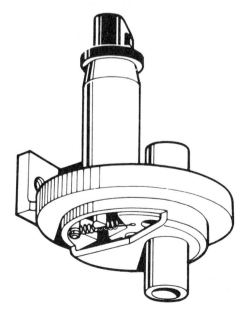

The auxiliary air regulator is another method used on fuel injection systems (see Air control valve) to provide extra air to the intake system during cold engine operation

Fuel Injection

Fuel distributor

The fuel distributor controls the amount of fuel passed into the injection system and distributes that fuel to the individual injectors. As the control plunger is moved in response to the air flow sensor, fuel passages are uncovered in the fuel distributor, allowing more or less fuel to pass through the distributor.

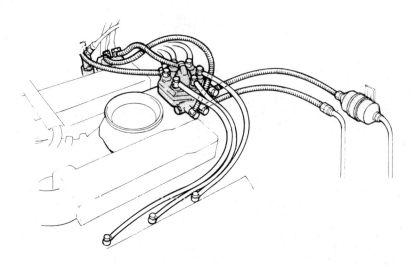

The fuel distributor is used on continuous flow injection systems to determine the amount of fuel delivered to the injectors, which are held open continuously while the engine is running by fuel pressure

Fuel accumulator

The fuel accumulator works in conjunction with the line pressure regulator to maintain the fuel delivery pressure in the fuel distributor by opening a valve which returns excess fuel to the tank when the pressure rises above that level.

Idle speed bypass valve

The idle speed on some systems is controlled by the ECM, through an idle air bypass valve. When the engine is cold or when accessory loads, such as air conditioning, necessitate a higher idle speed, the ECM, through an idle air control motor, opens the bypass valve to provide more air into the system, and at the same time to ECM increases the fuel injector open time to maintain the proper fuel/air ratio.

Applications 2

Listed below are the injection systems used on most 1978 and later model domestic and import automobiles. Keep in mind that in most cases the systems are the same, such as Bosch L-Jetronic fuel injection being used on Volkswagen and Nissan/Datsun, but the installation details differ from manufacturer to manufacturer.

Audi
    1984/later 4000S Quattro . . . . KE-Jetronic
    1978-79 Fox . . . . K-Jetronic
    1980/later 4000, Coupe . . . . K-Jetronic
    1978/later 5000, Turbo, Quattro . . . . K-Jetronic

BMW
    1978-82 320i . . . . K-Jetronic
    1983/later 318i . . . . L-Jetronic
    325e . . . . Motronic
    1979-81 528i . . . . L-Jetronic
    1982/later 528e . . . . Motronic
    1978 530i . . . . L-Jetronic
    1978-81 633CSi, 533i . . . . L-Jetronic
    1982/later 633CSi, 533i . . . . Motronic
    1978-81 733i . . . . L-Jetronic
    1982/later 733i . . . . Motronic

Fuel Injection

Buick
1982/later with 110 L4	TBI
1982/later with 110 L4 Turbo	MFI
1982/later with 122 L4	TBI
1982/later with 151 L4	TBI
1982/later with 231 V6	MFI
1985/later with 250 V8	DFI

Cadillac
1985/later with 231 V6	MFI
1981/later with 250 V8	DFI
1978-79 with 350 V8	EFI
1980-82 with 368 V8	DFI
1978-79 with 425 V8	EFI

Camaro
1982/later with 151 L4	TBI
1982/later with 305 V8	TBI

Chevrolet
1982/later with 110 L4	TBI
1982/later with 110 L4 Turbo	MFI
1982/later with 122 L4	TBI
1982/later with 151 L4	TBI
1982/later with 231 V6	MFI

Chrysler
1981-83 with 318 V8	TBI
1983/later with 135 L4	EFI

Application

Corvette
1982-84 with 350 V8 .... TBI
1985/later with 350 V8 .... MFI

Dodge
1981-83 with 318 V8 .... TBI
1983/later with 135 L4 .... EFI

Fiat/Bertone/Pininfarina
1980/later Spider 2000, Turbo .... L-Jetronic
1980-82 Strada .... L-Jetronic
1980/later X1/9 .... L-Jetronic
1980-81 Brava .... L-Jetronic

Ford
1983/later with 1597 L4 .... MPFI
1984/later with 140 L4 .... MPFI
1984/later with 232 V6 .... MPFI
1981/later with 302 V8 .... CFI

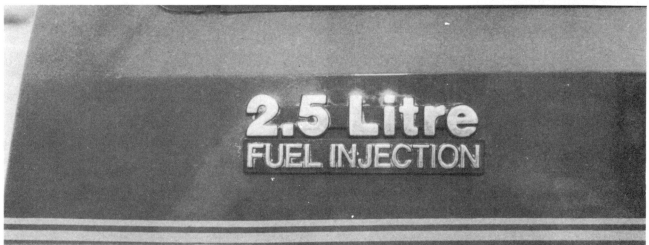

Isuzu
1983/later Impulse .... LH-Jetronic

Mercedes
1978-81 280E, SE, CE .... K-Jetronic
1978-80 450SL, 450SLC .... K-Jetronic
1978-80 450SEL .... K-Jetronic
1978-79 6.9 .... K-Jetronic
1981/later 380SL, SLC, SEL, SEC, SE .... K-Jetronic
1984/later 190E .... KE-Jetronic
1984/later 500SEC, SEL .... K-Jetronic

Mercury
1983/later with 1597 L4 .... MPFI
1984/later with 140 L4 .... MPFI
1984/later with 232 V6 .... MPFI
1981/later with 302 V8 .... CFI

Fuel Injection

Mitsubishi
1983/later Starion, Cordia, Tredia . . . . TBI

Nissan/Datsun
1984/later 300ZX, Turbo . . . . LH-Jetronic
1982/later 200SX . . . . L-Jetronic
1978-84 280Z, ZX . . . . L-Jetronic
1978/later 810 . . . . L-Jetronic

Oldsmobile
1982/later with 110 L4 . . . . TBI
1982/later with 110 L4 Turbo . . . . MFI
1982/later with 122 L4 . . . . TBI
1982/later with 151 L4 . . . . TBI
1982/later with 231 V6 . . . . MFI
1985/later with 250 V8 . . . . DFI

Application

Peugeot
1980/later 505 .... K-Jetronic

Plymouth
1981-83 with 318 V8 .... TBI
1983/later with 135 L4 .... EFI

Pontiac
1982/later with 110 L4 .... TBI
1982/later with 110 L4 Turbo .... MFI
1982/later with 122 L4 .... TBI
1982/later with 151 L4 .... TBI
1982/later with 231 V6 .... MFI

Porsche
1978/later 911SC .... K-Jetronic
1978-79 Turbo .... K-Jetronic
1978-81 924, 924 Turbo .... K-Jetronic
1978-79 928 .... K-Jetronic
1980/later 928 .... L-Jetronic
1982/later 944 .... Motronic

Renault
1982/later Alliance, Encore (California only) .... L-Jetronic
1982/later Alliance, Encore (except California) .... TBI
1981/later 18i, Sport Wagon .... L-Jetronic
1982/later Fuego .... L-Jetronic

Saab
1978 99, 99 Turbo .... K-Jetronic
1979/later 900, 900 Turbo .... K-Jetronic

Toyota
1979/later Supra, Celica .... L-Jetronic
1980/later Cressida .... L-Jetronic
1983/later Camry, Starlet .... L-Jetronic

Fuel Injection

Volkswagen
1978-79 Beetle (Type 1) .... L-Jetronic
1978/later Bus (Type 2), Vanagon .... L-Jetronic
1978-81 Dasher .... K-Jetronic
1980/later Jetta .... K-Jetronic
1978-83 Rabbit....................................... K-Jetronic
1984/later Rabbit .... KE-Jetronic
1978/later Scirocco .... K-Jetronic
1982/later Quantum .... K-Jetronic

Volvo
1978-79 242, 244, 245, 262, 264, 265 .... K-Jetronic
1980-83 DL, GL, GLE, GT, Coupe .... K-Jetronic
1984/later DL, GL, GLE, GT, Coupe .... LH II-Jetronic

Bosch Fuel Injection Systems

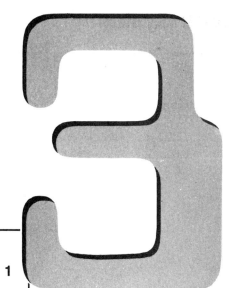

L-Jetronic System

L-Jetronic, LH-Jetronic and Motronic 1
General information
Troubleshooting

L/LH-Jetronic and Motronic components 2
Fuel pressure regulator
Electronic control unit
Air flow sensor
Air mass sensor
Throttle switch
Throttle vacuum switch
Oxygen sensor
Coolant temperature sensor
Fuel temperature sensor
Detonation sensor
Speed sensor
Boost sensor
Fuel injectors
Cold start valve
Thermo-time switch
Auxiliary air regulator
Fuel pump

Relieving L/LH-Jetronic and Motronic fuel system pressure 3

Typical L/LH-Jetronic and Motronic fuel system removal and installation procedures 4

K-Jetronic System

K-Jetronic and KE-Jetronic 5
General information
Troubleshooting

K/KE-Jetronic components 6
Electronic control unit
Air flow sensor
Throttle valve
Fuel distributor
Line pressure regulator
Fuel accumulator
Fuel injectors
Oxygen sensor
Cold start valve
Thermo-time switch
Warm-up regulator
Auxiliary air regulator
Fuel pump

Relieving K/KE-Jetronic fuel system pressure 7

Typical K/KE-Jetronic fuel system removal and installation procedures 8

Bosch Systems

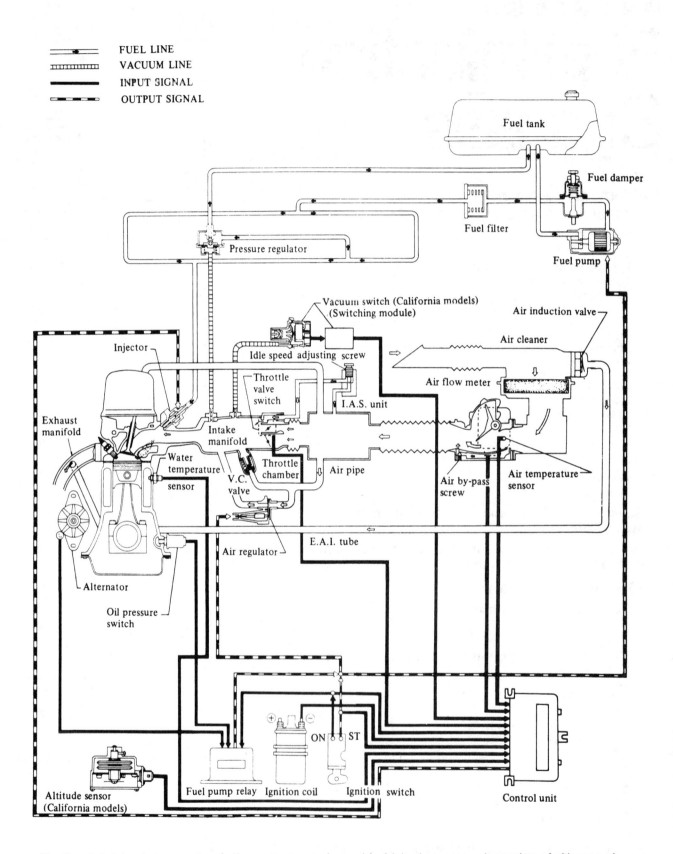

The Bosch L-Jetronic type system is the most commonly used fuel injection system. It consists of either an air flow or air mass sensor used to determine the amount of air entering the engine and electrically-operated injectors controlled by an Electronic Control Unit

1 L-Jetronic, LH-Jetronic and Motronic

The Bosch L-Jetronic fuel injection system is used by BMW, Fiat, Nissan (Datsun), Porsche, Renault, Toyota and Volkswagen.

The Bosch LH-Jetronic and Bosch Motronic fuel injection systems are used by Volvo, BMW and Porsche.

The systems are of the timed port-injection type, with injection solenoids delivering the fuel on demand, the length of time the solenoid is open determining the amount of fuel delivered to the engine.

The time the injector is held open, and therefore the amount of fuel that is delivered into the engine, is determined by the Electronic Control Unit (ECU), which receives signals from a variety of sensors, the key unit of which is the air flow meter, which measures the amount of air entering the induction system.

The LH-Jetronic and Motronic systems measure air mass, rather than flow.

Measurement of the air mass is accomplished by using a very thin platinum wire in the air mass sensor unit. Current is passed through the wire, heating it. Air flowing over the wire draws heat away from it, which changes the electrical resistance of the wire. In response to the change in resistance, the electronic amplifier unit increases the current to keep the wire at a constant temperature (and resistance), and the ECU interprets the changes in current as changes in air mass flow, which determines the amount of time the injectors are held open.

General information

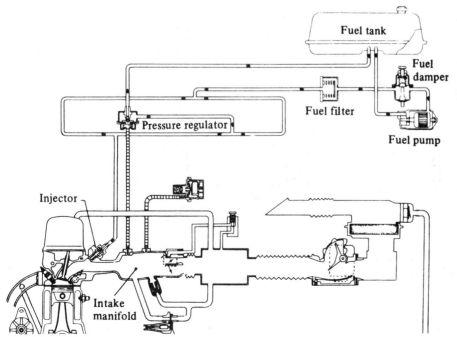

Fuel Flow

Fuel is delivered to the fuel pressure regulator by an electric fuel pump. The regulator maintains a constant pressure differential between the fuel system and the intake manifold. Pulsations in the fuel system caused by the opening and closing of the injectors are eliminated by the fuel damper

Bosch Systems

The ECU also receives signals from a throttle switch which tells the ECU whether the throttle is closed (idle) or wide-open (high demand), an oxygen sensor which tells the ECU if the mixture is too rich or too lean, a coolant temperature sender to tell the ECU if the engine is still cold, requiring more fuel for warm up, a fuel temperature sender, a detonation sensor, a vehicle speed sensor and a boost sensor for turbocharged engines. Not all of these sensors are used on all engines, but the basic design and operation of the L-Jetronic system is the same for all applications.

Air Flow

The air flow into the engine is measured by an air flow meter, which signals the Electronic Control Unit to control the amount of time the injectors are held open. A throttle valve controls the amount of air passed into the intake manifold, and the air regulator bypasses air around the throttle valve to increase the idle speed during cold engine operation

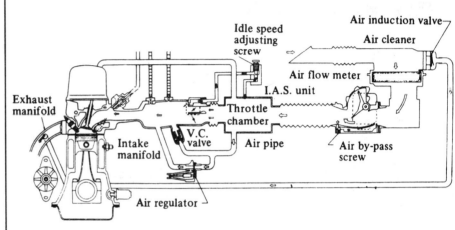

Electrical system

As the throttle valve is opened, more air passes through the air flow meter, and an electrical signal is sent to the Electronic Control Unit. Since the fuel pressure regulator maintains a constant pressure differential between the fuel system and the intake manifold, an electrical signal sent to the fuel injector to hold it open for a specific amount of time controls precisely the amount of fuel injected. Electrical signals from a variety of sensors are used to modify the length of the injector pulse signalled by the control unit

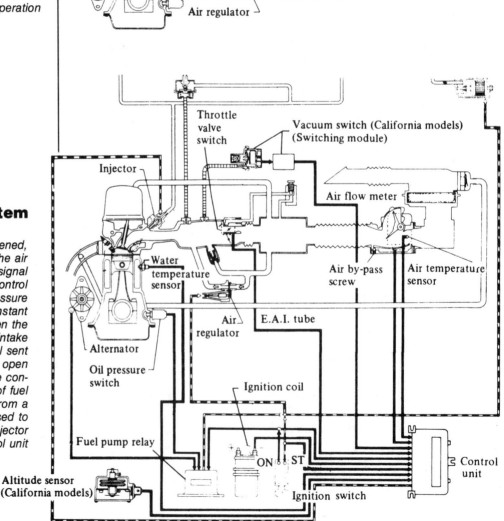

L-Jetronic

Troubleshooting

Note: All diagnostic causes listed below relate only to the fuel injection system. Quite often the symptom described may have causes other than in the fuel injection system. As an example, "Engine will not start" may well be due to a fault in the ignition system, although such causes are not listed here.

Engine will not start:
Vacuum leak
Leak in air intake system
Fuel line or filter blocked
Fuel pump not working
System fuel pressure incorrect
Cold start valve inoperative
Cold start valve leaking
Thermo-time switch defective
Auxiliary air valve malfunctioning
Temperature sensor defective
Air flow meter not functioning properly
Control unit defective

Engine starts then dies:
Leak in air intake system
Fuel line or filter blocked
System fuel pressure incorrect
Cold start valve leaking
Auxiliary air valve malfunctioning
Temperature sensor defective
Air flow meter not functioning properly
Idle speed setting incorrect
CO setting incorrect
Control unit defective

Uneven idle:
Leak in air intake system
Fuel line or filter blocked
System fuel pressure incorrect
Cold start valve leaking
Auxiliary air valve malfunctioning
Temperature sensor defective
Air flow meter not functioning properly
Throttle plate not closing properly
Throttle position indicator switch defective
Idle speed setting incorrect
Injector defective
CO setting incorrect
Control unit defective

Idle speed incorrect:
Leak in air intake system
Cold start valve leaking
Auxiliary air valve malfunctioning
Air flow meter not functioning properly
Throttle plate not closing properly
Idle speed setting incorrect

Bosch Systems

Troubleshooting
(continued)

Engine will not maintain an even speed:
- Leak in air intake system
- System fuel pressure incorrect
- Cold start valve leaking
- Air flow meter not functioning properly
- Injector defective
- CO setting incorrect

Engine misses:
- Leak in air intake system
- Injector defective
- Control unit defective

Excessive fuel consumption:
- System fuel pressure incorrect

- Cold start valve leaking
- Temperature sensor defective
- Air flow meter not functioning properly
- CO setting incorrect

Loss of power:
- Leak in air intake system
- Fuel line or filter blocked
- System fuel pressure incorrect
- Air flow meter not functioning properly
- Throttle plate not opening completely
- Throttle position indicator switch defective
- Injector defective
- Control unit defective

2 L/LH-Jetronic and Motronic components

Fuel pressure regulator

A key component of the system is the fuel pressure regulator, which works with the fuel pump to maintain a steady pressure relationship between the fuel line side of the injectors and the intake manifold.

Since the fuel is metered by the ECU by altering the amount of time the injector is held open, the volume of fuel injected would vary if there was a pressure difference between the fuel supply and the intake manifold under different conditions. When there is a high vacuum in the intake manifold, such as at high rpm, the pressure in the fuel system must be reduced. When there is low pressure in the intake manifold, such as at low speed, full throttle operation, the pressure in the fuel line must be increased.

The essential requirement is that the pressure differential remain the same under all circumstances, so that only the time the injector is open determines the amount of fuel injected. In order to maintain the desired fuel pressure, excess fuel is returned to the fuel tank by a separate line from the fuel pressure regulator.

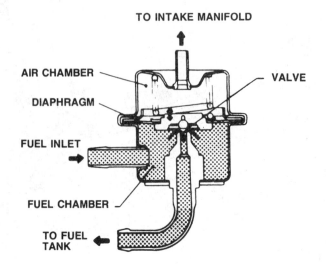

A key unit of the electronically-controlled fuel injection system is the fuel pressure regulator, which monitors the pressure in the intake manifold and adjusts the pressure in the fuel system so that there is a constant differential in pressure between the two areas. Because the pressure differential remains constant, the length of time the injector is held open controls precisely the amount of fuel injected into the engine

L-Jetronic

Electronic control unit

The Electronic Control Unit (ECU) receives the signals from the sensors associated with the fuel injection unit, such as the throttle position sensor, air flow sensor, etc., and sends out the signals to the injectors to control the amount of fuel injected into the cylinders.

Integrated into the ECU is the Programmable Read Only Memory unit (PROM), which contains the instructions for that particular vehicle and fuel injection design. Each manufacturer has different operating parameters for the fuel injection system, and therefore each has different PROMS. Because of this, the ECU cannot be interchanged between different makes of vehicles, even when those makes all use L-Jetronic, LH-Jetronic or Motronic systems, and in many cases the ECU cannot even be interchanged among different models of the same manufacturer.

The ECU used with the Motronic system contains additional capabilities not used in the L-Jetronic and LH-Jetronic systems, primarily a system to constantly monitor and adjust the ignition timing for optimum engine performance.

The ECU is a solid state device (no moving parts), and very seldom will an ECU fail in service. This is not to say, however, that an ECU cannot go "bad." But when an ECU does fail, it is almost invariably because something outside the unit has failed, interferring with the signals into our out of the ECU. In this case, it is common for the ECU to self-destruct, and since it is a very expensive piece of equipment, it is advisable to exercise extreme caution when working with any components connected to the ECM.

The two key precautions to take when working with the ECU or ECU-connected components are: 1) never disconnect the ECU with the ignition switch in the On position, and 2) always take extreme care with connectors to make sure that connecting pins enter only the correct receptacles.

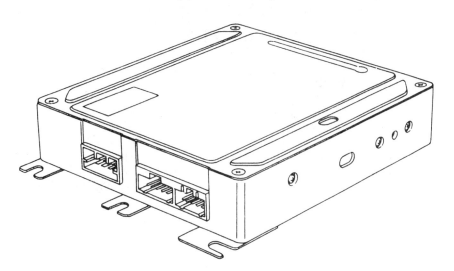

The Electronic Control Unit (also called the Electronic Control Module) is the heart of the fuel injection system. It receives signals from various sensors and uses the data received to determine the amount of fuel needed at any given moment

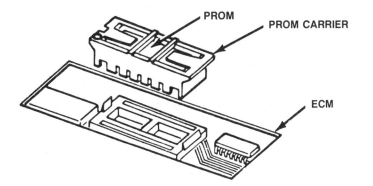

A Programmable Read Only Memory (PROM) is integrated into the Electronic Control Unit, and it is the PROM which contains the instructions which control the fuel injection system (and usually ignition timing) for your particular make and model vehicle

Bosch Systems

Air flow sensor

Installed in the air intake passage of the L-Jetronic system is a movable flap connected to a sensor which measures the amount of air being inducted into the system. The more air entering the system, the farther the flap is forced open.

At the same time the signal is being sent by the air flow sensor, other signals are coming into the ECU from the throttle position sensor, the engine rpm sensor and the air temperature sensor. All of these signals are integrated by the electronic control unit to determine how long the injectors should be held open.

In addition to the flap of the air flow sensor, there is a small bypass port which controls the idle speed by increasing (or decreasing) the amount of air which passes through the air flow sensor at idle.

Because the measurement of the amount of air flowing into the engine is the primary determinant of the amount of fuel injected, it is critical that only air passing through the flow meter be allowed into the engine. Any leaks in the system will not be accounted for by the air flow meter, which can cause the engine to run lean (or not run at all).

The air flow sensor measures the amount of air entering the engine and sends a signal to the Electronic Control unit, which opens the injectors just long enough to provide the proper amount of fuel to mix with the incoming air

Air mass sensor

Installed in the air intake passage of the LH-Jetronic and Motronic systems is a platinum wire which is heated by a current from an electronic amplifier, which measures the resistance to the current flow. The hotter the wire becomes, the greater the resistance. The amplifier controls the current flow so as to maintain the wire at an even temperature, and therefore at an even level of resistance.

Air flowing past the wire cools it, and the more air flowing past the wire, the greater the cooling effect. As the wire cools, the amplifier has to send more current through the wire to keep it at the set temperature, and this greater current flow is signalled to the ECU, which injects more fuel to match the greater air flow.

At the same time the signal is being sent by the air mass sensor, other signals are coming into the electronic control unit from the throttle position sensor, the engine rpm sensor and the air temperature sender. All of these signals are integrated by the ECU to determine how long the injectors should

L-Jetronic

be held open, and with the Motronic system the signals are also used to adjust the ignition timing (advance) for optimum engine operation.

Because the measurement of the amount of air flowing into the engine is the primary determinant of the amount of fuel injected, it is critical that only air passing through the flow meter be allowed into the engine. Any leaks in the system will not be accounted for by the air flow meter, which can cause the engine to run lean (or not run at all).

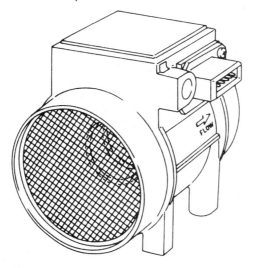

An air mass sensor also measures the incoming air charge, but instead of moving a vane, the incoming air cools an electrically heated wire, and the more air passing through the meter, the greater the cooling action. The Electronic Control Unit interprets this cooling as air mass flow, and adjusts the amount of fuel injected accordingly

Throttle switch

The throttle switch is mounted on the throttle chamber, moves in accordance with the movement of the accelerator pedal, and sends a position signal to the ECU. In most applications the switch monitors idle and wide-open throttle positions, although some have a third position, signalling mid-range throttle opening.

When the throttle is in the idle position, the throttle switch signals the ECU that idle enrichment is needed, and, when the ignition switch is shut off, the throttle switch sends the signal which shuts off the fuel flow from the tank by cutting the electrical power to the fuel pump. The idle switch also cancels the input to the oxygen sensor during coast-down to keep the system from being affected by the excessive oxygen in the system at that time.

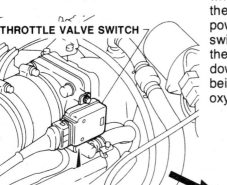

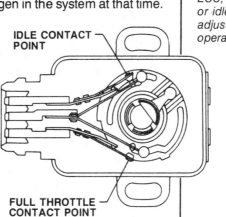

The throttle switch (also called the throttle position switch) is mounted on the throttle valve body, where it signals idle and full throttle conditions to the Electronic Control Unit. The ECU, when it receives a full throttle or idle signal from the throttle switch, adjusts the fuel/air mixture for these operating conditions

Bosch Systems

Throttle vacuum switch

On some models the throttle switch is replaced with a vacuum switch, which indicates both full throttle and idle by low ported vacuum and part throttle operation by high manifold vacuum.

Oxygen sensor

The oxygen sensor is installed in the exhaust manifold. It measures the amount of unburned oxygen remaining in the exhaust gas and signals the ECU to change the amount of fuel being injected as necessary to maintain the proper fuel/air mixture.

There are two times when the oxygen sensor is cut out of the circuit. When the throttle is closed (at idle), since the best idle mixture, both from an emissions and smoothness standpoint, is leaner than a "running" mixture, the throttle position sensor cancels the signal being received from the oxygen sensor at the ECU.

A second time when the oxygen sensor signal to the ECU is diverted (also by the throttle position sensor) is under full throttle conditions, when a slightly over-rich mixture is required to keep exhaust gas temperatures down, protecting the catalytic converter and oxygen sensor.

The oxygen sensor operates under very low voltage conditions, and therefore current, either from the battery or from a test instrument, should never be applied to the oxygen sensor.

The oxygen sensor is mounted in the exhaust manifold or high in the exhaust pipe, where it monitors the exhaust gases, sending a signal to the Electronic Control Unit, which adjusts the fuel/air mixture to the proper ratio

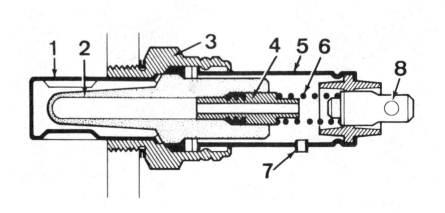

The oxygen sensor is a delicate instrument, and care should be taken when removing and installing it to prevent damage to the internal components

1 Protective outer sleeve
2 Ceramic sensor unit
3 Base
4 Contact cover
5 Protective cover
6 Contact spring
7 Ambient air inlet
8 Electrical connector

L-Jetronic

The oxygen sensor does not function until it reaches a rather high operating temperature, so on some later models the oxygen sensor includes a heater unit to bring the sensor into operation as quickly as possible. The later model heated sensor can be identified by noting that it has three wires, rather than two, in the connector. The early (unheated) and late (heated) sensors cannot be interchanged.

Coolant temperature sensor

When the engine is cold, additional fuel is required, and, just as the choke on a carburetor enrichens the mixture by reducing the amount of air taken into the engine, the coolant temperature sensor signals the ECU to hold the injectors open longer when the engine is cold, passing more fuel into the engine.

Two methods are used to signal the engine temperature to the ECU. On many engines a temperature sensor is inserted into the cooling system, much like the coolant temperature probe which activates the temperature gauge or warning light. As the coolant warms the extra enrichment is reduced, until the engine is at normal operating temperature.

On some engines, especially later models and those with turbochargers, a cylinder head temperature sensor is used instead of the coolant sensor. In this case the actual temperature of the engine is signalled to the ECU, which gives a somewhat more accurate indication of operating conditions than the coolant temperature sensor.

The coolant temperature sensor monitors the temperature in the cooling system, signalling the Electronic Control Unit when extra air and fuel is necessary for cold engine operating conditions

Fuel temperature sensor

As gasoline heats, it expands, and therefore warmer fuel would result in a leaner mixture. Because of this, L-Jetronic units incorporate a fuel temperature sensor, generally built into the fuel pressure regulator, which signals the ECU to enrichen the mixture whenever the fuel temperature exceeds a certain value.

A fuel temperature sensor mounted in the pressure regulator monitors the temperature of the fuel and, when the temperature exceeds a predetermined level the Electronic Control Unit richens the mixture to compensate

Bosch Systems

Detonation sensor

On some engines, generally those installed in high performance models and those equipped with a turbocharger, a detonation sensor is included in the fuel injection sensor system to detect engine knock, caused by preignition of fuel with too low an octane rating or excessive cylinder temperatures, usually caused by excessively advanced ignition timing.

When detonation is detected by the vibration sensor, which is usually installed in the cylinder block, a signal is sent to the ECU to retard the ignition timing, richen the mixture, or both.

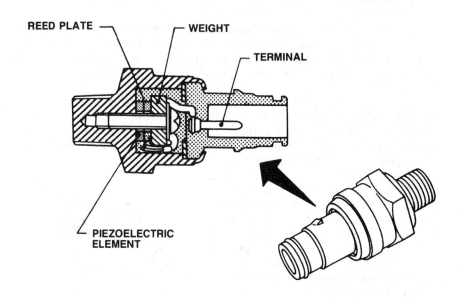

To prevent internal engine damage from detonation, a sensor signals the Electronic Control Unit, which richens the mixture (and on some systems retards the ignition timing) to eliminate the conditions causing the detonation

Speed sensor

A road speed sensor is included with some systems to give the electronic control unit a modifying signal for the engine rpm sensor. The sensor is mounted in the speedometer unit and is made up of a reed-type switch on analog speedometers or a LED, photo diode and shutter on digital type speedometers.

Boost sensor

On turbocharged models a boost sensor is incorporated in the system. The sensor signals the ECU under high boost conditions to enrichen the mixture. On some models the ECU also controls the wastegate to limit boost and cuts off the fuel delivery to the injectors to prevent engine damage.

Fuel injectors

The fuel injector is an electrical solenoid, controlled by the ECU. Fuel under pressure is supplied to the injectors, and a pressure differential of approximately 36 psi is maintained between the fuel line pressure on one side of the injector and the manifold pressure on the other side.

Inside the injector unit is a coil. When current is supplied to the coil the injector valve opens, allowing fuel to pass through the unit and into the intake manifold. The ECU controls the current reaching the injector. The longer the ECU sends an "open" signal, the more fuel is injected into the engine.

Although there is one injector for each cylinder, mounted in the intake manifold upstream from each intake valve, all the injectors receive the signal from the ECU at the same time, so the actual injection of fuel into the intake passage is independent of the position of the piston (on the intake stroke,

L-Jetronic

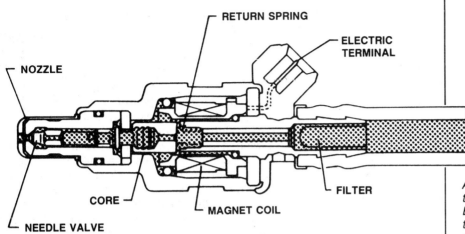

A coil inside the fuel injector opens the needle valve at a signal from the Electronic Control Unit, allowing fuel to be injected into the engine. The amount of time the injector is held open (called the pulse width) determines the fuel/air mixture ratio

exhaust stroke, etc.) at the time of injection. When the intake valve opens, the fuel has been injected into the port and is ready to be drawn into the cylinder.

With the intake valve opening to draw in fuel some 30 times per second or more at "cruising" rpm for most engines (approximately 3500 rpm), the fact that the fuel is not injected at exactly the moment that the intake valve begins to open is unimportant.

Cold start valve

Some applications utilize a cold start valve to make it easier to start a cold engine. The cold start valve is actually an additional injector solenoid, usually mounted centrally in the intake manifold or plenum area.

The cold start valve is energized through the starting system, spraying extra fuel into the manifold while the starter is turning the engine over. To prevent excessive fuel from being injected into the engine during starting, a thermo-time switch is incorporated in the circuit.

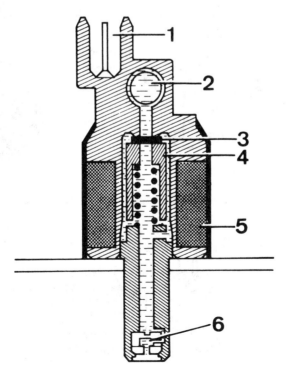

The cold start valve is a fuel injector located near the throttle plate assembly which is used to supplement the fuel delivered by the injectors during cold-engine operation
 1 Electrical connector
 2 Fuel intake
 3 Seal
 4 Solenoid
 5 Coil winding
 6 Nozzle

Bosch Systems

Thermo-time switch

The thermo-time switch controls the amount of time the cold start valve remains operational. The temperature sensor part of the switch monitors the coolant temperature, while a timer inside the switch ensures that the cold start valve does not operate for longer than a set period of time

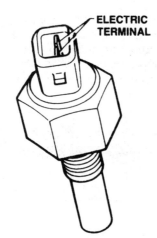

The thermo-time switch regulates the amount of time the cold start valve is energized and prevents the opening of the cold start valve when the engine is warm.

The "thermo" section of the switch is a heater coil. When the starter is energized, current is passed through the thermo-time switch and into the cold start valve. As the current passes through the switch, the heater coil warms a bimetallic switch. It takes between 8 and 12 seconds for the heater coil to warm the bimetallic switch to the point where it opens the circuit, preventing further fuel from passing through the cold start valve.

The length of time the circuit is energized (8 to 12 seconds) is controlled by the coolant temperature. The lower the coolant temperature, the longer it will take the heater coil to warm the bimetallic switch.

Auxiliary air regulator

The auxiliary air regulator bypasses extra air around the throttle valve when the engine is cold to increase the idle speed. An electric heater inside the regulator warms a bimetal strip, which slowly closes a shutter until, when the engine is warmed, all the air is being routed through the throttle valve

The auxiliary air regulator acts as the "choke" for the fuel injection system. When the engine is cold, it bypasses extra air around the throttle plate, which is read as extra air flow by the ECU, which signals for additional fuel.

A bimetallic spring, activated by coolant temperature, and an electrical heater coil, to prevent excessive operation of the auxiliary air regulator, control a rotating disc inside the throttle chamber. The disc has a hole in it, and when the hole lines up with a matching hole in the throttle chamber bypass, extra air is inducted into the engine. As the bimetallic spring heats up, it rotates the disc, gradually closing off the hole until all the air is passing into the engine through the main throttle chamber passage.

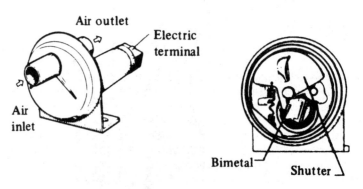

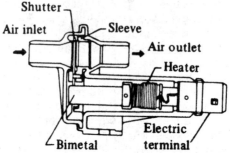

L-Jetronic

Fuel Pump

The fuel pumps used with L-Jetronic systems are generally of the wet vane type, where the pump is mounted in the fuel tank and is filled with fuel. The pump is both cooled and lubricated by the fuel, and therefore the pump should never be allowed to run dry. Operating the pump for more than a few seconds without fuel in the pump for cooling and lubrication can cause the pump to seize.

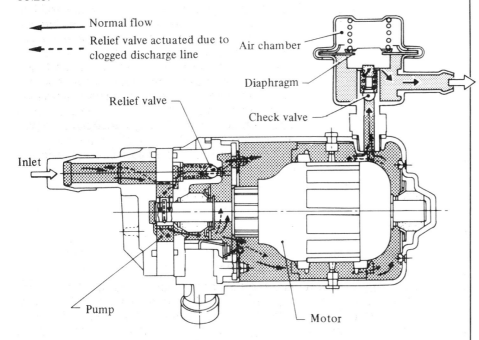

The electric fuel pumps used on fuel injection systems are usually of the wet vane type, where fuel surrounds the motor and acts as a cooling agent and lubricant. It is important that this type of pump never be run when the fuel tank is empty or the fuel line from the tank is disconnected

3 Relieving L/LH-Jetronic and Motronic fuel system pressure

The fuel lines are under high pressure (35 to 45 psi), and a check valve closes to maintain the pressure in the lines even after the engine has been shut off. The pressure can remain in the lines even after the vehicle has been out of use for several days. Because of this, it is essential that the pressure in the lines be relieved before disconnecting any fuel system components.

There are several methods for relieving pressure in the fuel system. The accessibility of components on your particular vehicle should determine which method is used.

The simplest method, most often usable on early L-Jetronic systems, is to start the engine and disconnect either the fuel pump relay or the fuel pump electrical connector. Allow the engine to continue running until it stops from lack of fuel. Disconnect the negative battery cable, then reconnect the relay or fuel pump connector.

A second method utilizes the cold start valve to remove the pressure in the fuel lines. Start by removing the ground cable from the battery, then disconnect the cold start valve wiring connector. Use two jumper wires to connect the battery directly to the cold start valve electrical terminals. Use extreme caution here as these terminals are very close together and carelessness could easily lead to a direct short. It will only be necessary to

Bosch Systems

connect the wires for two or three seconds to relieve the pressure in the fuel system.

The final method for relieving fuel system pressure requires the use of a hand-operated vacuum pump. Disconnect the vacuum line (not the fuel line!) from the fuel pressure regulator and attach the hand vacuum pump to the pressure regulator vacuum line fitting. Operate the hand pump until there is 20 in-Hg of vacuum applied to the pressure regulator, which will open the regulator return line to the tank, releasing the pressure in the system.

4 Typical L/LH-Jetronic and Motronic fuel system removal and installation procedures

Note: While installation on various makes and models of vehicles varies due to manufacturer installed component location, engine configuration (V-type, inline, flat opposed) and number of cylinders, the basic design of the L-Jetronic, LH-Jetronic and Motronic systems will be the same in all cases, and the following typical removal and installation procedure will, with only slight modifications, outline the procedure for your vehicle.

Removal

1 Relieve the pressure in the fuel system.
2 Disconnect the ground cable at the battery.
3 Disconnect the air ducts and hoses connecting the air cleaner and the air flow meter.
4 Remove the air cleaner-to-mounting bracket bolts.
5 Disconnect the air flow meter electrical connector.
6 Remove the bolts connecting the air flow meter to the air cleaner housing.
7 Remove the air flow meter.
8 Remove the air cleaner housing.
9 Remove the air duct to the throttle valve housing.
10 Disconnect the throttle position sensor electrical connector.

Large clamps are used to seal the ducting between the air cleaner and air flow meter

The throttle position sensor is mounted on the side of the throttle valve housing

L-Jetronic

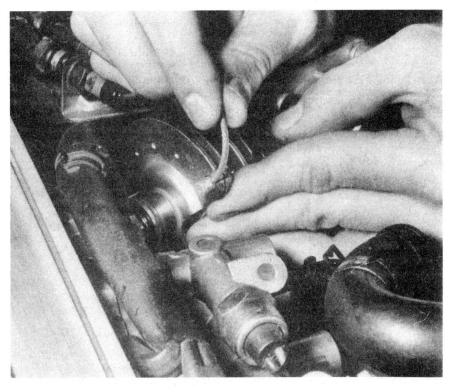

Disconnect the throttle cable from the throttle valve, being careful to note how it is installed so that it can be properly reinstalled

Disconnect any vacuum lines, being careful to label them for proper reconnection, then remove the throttle valve housing from the intake manifold

11 Remove the throttle cable from the throttle valve housing.

12 Remove the throttle valve housing from the intake manifold.

13 Remove the dash pot from the throttle valve housing.

14 Remove the screws securing the throttle position sensor to the throttle valve housing.

Bosch Systems

Removal
(continued)

15 Carefully remove the throttle position sensor from the throttle valve housing.
16 Drain the cooling system.
17 Disconnect the coolant temperature sensor electrical connector.
18 Remove the coolant temperature sensor.
19 Disconnect the oxygen sensor electrical connector.
20 Remove the oxygen sensor from the exhaust manifold.
21 Disconnect the injector harness electrical connector.
22 Disconnect the air regulator hose.
23 Remove the air regulator.
24 Disconnect the air regulator electrical connector.
25 Disconnect the fuel hose from the fuel rail.
26 Disconnect the fuel return line from the fuel rail.
27 Disconnect the vacuum line from the fuel rail.
28 Remove the vacuum hose connecting the fuel pressure regulator to the intake manifold.
29 Remove the bolts securing the fuel rail to the intake manifold.

Remove the two screws holding the throttle position sensor to the throttle valve housing and remove the throttle position sensor

Release the clamps and disconnect the hoses from the auxiliary air regulator, release the electrical connector and remove the air regulator

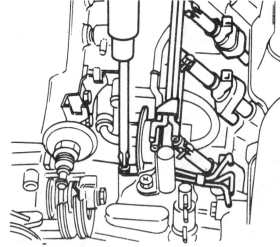

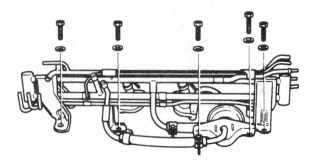

Remove the screws holding the fuel rail to the intake manifold, being careful not to lose the washers

L-Jetronic

Two screws attach each injector to the intake manifold

Lift off the fuel rail, fuel pressure regulator and the injectors as an assembly

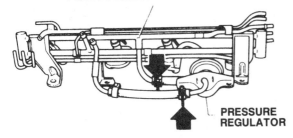

Loosen the clamps (arrows) and remove the fuel pressure regulator from the fuel rail

30 Remove the screws securing the fuel injectors to the intake manifold.

31 Remove the fuel rail, fuel pressure regulator and fuel injectors from the intake manifold as a unit.

32 Release the hose clamps on the fuel injectors and pull the injectors from the fuel rail.

33 Release the clamps and remove the fuel pressure regulator from the fuel rail.

Loosen the clamps and remove each injector from the fuel rail

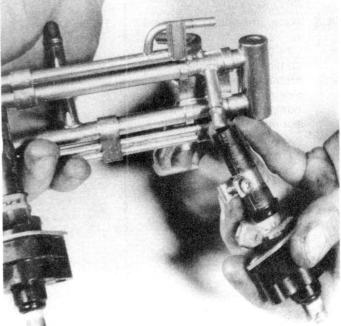

47

Bosch Systems

Installation

34 Install the fuel pressure regulator on the fuel rail.

35 Install the fuel rail, fuel pressure regulator and fuel injectors as a unit by inserting the fuel injectors into the openings in the intake manifold. Use light oil on the O-rings and make sure they are seated properly as the injectors are installed.

36 Install the screws which hold the fuel injectors in place in the intake manifold.

37 Install the fuel rail retaining bolts.

38 Connect the fuel pressure regulator to intake manifold vacuum hose.

39 Connect the vacuum line to the fuel rail.

40 Connect the fuel line and fuel return line to the fuel rail.

41 Connect the air regulator electrical connector.

42 Install the air regulator hose.

43 Connect the injector harness to the main harness.

44 Apply anti-seize compound to the threads of the oxygen sensor and install the sensor in the exhaust manifold. Use caution to ensure that none of the anti-seize compound gets on the sensor end.

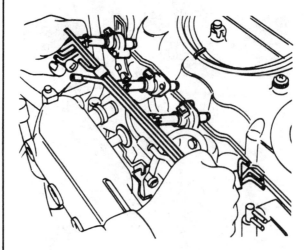

Reinstall the fuel injectors, fuel rail and fuel pressure regulator onto the intake manifold as a unit

Install the two screws which hold each fuel injector to the intake manifold

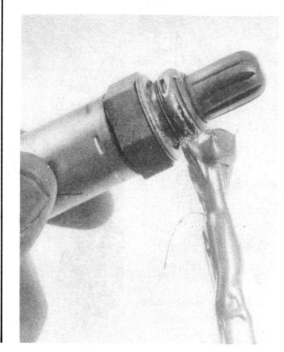

Before installing the oxygen sensor in the exhaust manifold the threads must be coated with anti-seize compound. Use caution to ensure that none of the compound gets on the sensor end

L-Jetronic

When setting the throttle position sensor, rotate the sensor in a clockwise direction until the ohmmeter indicates no continuity in the circuit

Install the dashpot on the throttle valve housing

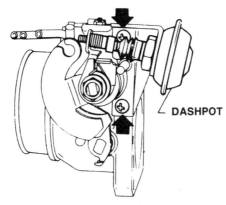

45 Connect the oxygen sensor electrical connector.

46 Install the coolant temperature sensor. Be sure the copper gasket is in place between the sensor and the engine.

47 Install the throttle position sensor on the throttle valve housing but do not tighten the bolts.

48 Connect an ohmmeter to the two top electrical terminals on the throttle position sensor.

49 Rotate the throttle position sensor clockwise until the ohmmeter shows no continuity in the circuit.

50 Turn the throttle position sensor back counterclockwise until continuity is shown on the ohmmeter and tighten the bolts.

51 Install the dash pot on the throttle valve housing.

52 Install the throttle valve housing on the intake manifold.

53 Install the throttle cable.

54 Connect the throttle position sensor electrical connector.

55 Install the air duct between the air cleaner and throttle valve housing.

56 Install the air cleaner assembly.

57 Install the air flow meter in the air cleaner assembly.

58 Connect the air flow meter electrical connector.

59 Install the air cleaner bracket bolts.

60 Replace all removed ducts and hoses.

61 Fill the cooling system.

62 Connect the ground cable to the battery.

Bosch Systems

5 K-Jetronic and KE-Jetronic

General information

The Bosch K-Jetronic system is used on Audi, BMW, Mercedes, Peugeot, Porsche, SAAB, Volkswagen and Volvo. It is a Constant Injection System (CIS), in which the fuel flows through the fuel injectors as long as the engine is running, the mixture being controlled by the fuel pressure through the injectors, which are spring loaded and calibrated to open only when the pressure exceeds a certain point, rather than by the amount of time the injector is open.

Two types of K-Jetronic systems have been used. Before 1983 the K-Jetronic was a completely mechanical system, with no electronic computer controls. After 1983 an electronic control system was integrated into the system to control engine functions.

There is one injector nozzle for each cylinder. The injectors are spring-loaded and held closed until the fuel pressure exceeds a certain level. There are no electrical components in the injectors as there are with the other Bosch injector systems.

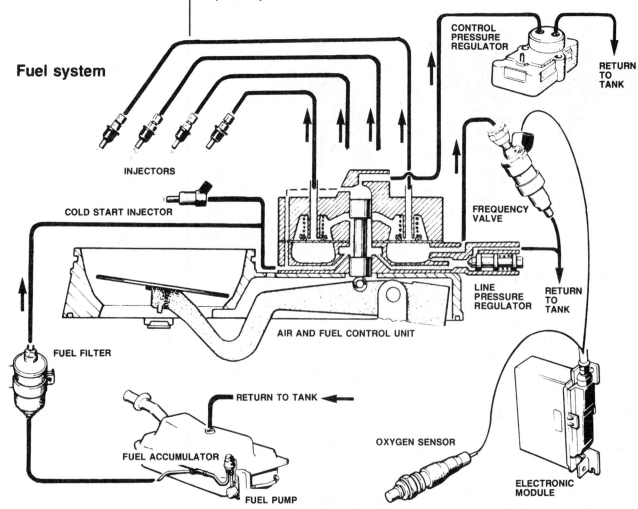

The K-Jetronic system utilizes a fuel distributor to control the amount of fuel injected into the system and injectors which remain open whenever the engine is running

K-Jetronic

The fuel pump delivers fuel to the fuel distributor, which includes a line pressure regulator, a control plunger and a pressure regulator valve for each injector. The line pressure regulator ensures that fuel at a constant pressure is delivered to the control plunger and the plunger controls the amount of fuel which is actually delivered to the pressure regulator valves.

The air/fuel control unit contains an air flow sensor, which is mechanically connected to the control plunger in the fuel distributor. The greater the volume of air moving through the air/fuel control unit, the more the air flow sensor causes the control plunger to be moved, making more fuel available to the pressure regulator valves.

Troubleshooting

Note: All diagnostic causes listed below relate only to the fuel injection system. Quite often the symptom described may have causes other than in the fuel injection system. As an example, "Engine will not start" may well be due to a fault in the ignition system, although such causes are not listed here.

Engine will not start:
Vacuum leak
Air flow sensor plate binding
Auxiliary air valve not opening
Fuel pump inoperative
Cold start valve not working
Cold start system not energized
Cold control pressure incorrect

Engine hard to start when cold:
Vacuum leak
Air flow sensor plate or control plunger sticking
Inoperative auxiliary air valve
Fuel pump not working
Cold start system or valve inoperative
Cold control pressure incorrect

Uneven idle during warm-up:
Vacuum leak
Air flow sensor plate binding
Incorrectly adjusted air flow sensor plate stop
Fuel pump not working properly
Leaking cold start valve
Incorrect cold control pressure
Injectors leaking
Unequal fuel delivery between cylinders
CO adjustment incorrect
Idle speed adjustment incorrect

Engine backfires:
Warm control pressure incorrect
CO adjustment incorrect
Idle speed adjustment incorrect

Bosch Systems

Troubleshooting
(continued)

Uneven idle when engine is warm:
Vacuum leak
Air flow sensor plate binding
Leaking cold start valve
Warm control pressure incorrect
Injectors leaking
CO adjustment incorrect
Idle speed adjustment incorrect

Engine misfires under load:
Vacuum leak
Fuel pump not operating correctly
Warm control pressure too high
System fuel pressure incorrect
Injectors leaking

Loss of power:
Air flow sensor plate binding
Warm control pressure incorrect
System fuel pressure incorrect
Unequal fuel delivery between cylinders
Throttle plate not opening fully

Engine runs after switching off (diesels):
Air flow sensor plate binding
Air flow sensor plate stop not set correctly
Leaking cold start valve
Injectors leaking
CO adjustment incorrect
Idle speed setting incorrect

Excessive fuel consumption:
Leaking cold start valve
Warm control pressure too low
CO adjustment incorrect
Idle speed setting incorrect

Flat spot during acceleration:
Vacuum leak
Air flow sensor plate binding
Warm control pressure incorrect
System fuel pressure incorrect
Unequal fuel delivery between cylinders
CO adjustment incorrect
Idle speed setting incorrect

6 K/KE-Jetronic components

Electronic control unit

The early K-Jetronic unit was completely mechanical, the air flow through the throttle chamber acting on the air flow sensor plate, which is connected directly to the fuel control plunger. The later model K-Jetronic systems incorporate an ECU which receives signals from the coil (engine rpm), coolant temperature sensor, airflow sensor, throttle valve switch and oxygen sensor.

The ECU uses these signals to modify both the fuel control plunger and the idle speed. In addition, some models contain an altitude compensator, which adjusts the fuel mixture to compensate for changes of altitude (air density).

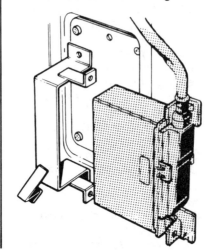

The electronic module receives a signal from the oxygen sensor to determine whether the mixture is too rich or too lean, and an appropriate adjusting signal is sent to the frequency valve to increase or decrease the fuel pressure in the fuel distributor

K-Jetronic

Air flow sensor

The air flow sensor is a precisely counterbalanced plate located in the air venturi. As the throttle is opened, air flowing through the venturi depresses the air flow sensor plate. The more air there is flowing into the engine, the more the plate is depressed.

The plate, in turn, is connected to the control plunger in the fuel distributor, which regulates the amount of fuel being delivered to the injectors.

Because the amount of fuel delivered to the injectors, and therefore the fuel/air ratio, is directly controlled by the movement of the air flow sensor plate, the system is extremely sensitive to air leaks, which will cause the engine to run lean.

The rubber connection between the air flow sensor and the throttle valve housing, the cold start valve gasket, the throttle valve to intake manifold gasket and the crankcase ventilation hose from the oil filler cap are areas which are susceptible to air leakage. Other areas which should be checked if leaks are suspected are the oil dipstick, valve cover gasket and all vacuum hoses.

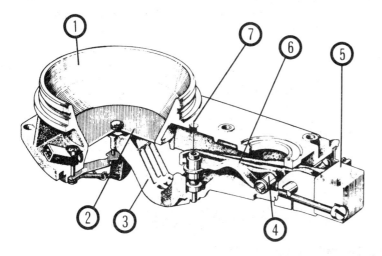

The air flow sensor measures the amount of air flowing into the engine through the displacement of the sensor plate (2), which opens or closes the venturi (1). The plate is attached to a lever (3), which is mounted on a flange (4), and is counterbalanced by a counterweight (5). The adjustment screw (6) attached to the lever (6) is used to adjust the CO content

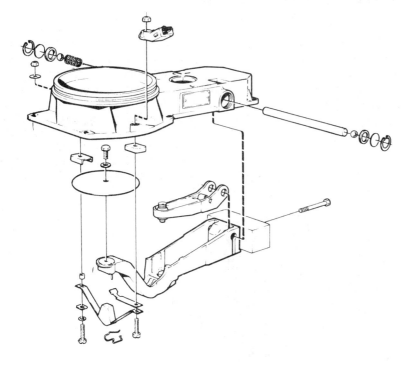

This exploded view of the air flow sensor shows the relationship of the counterbalanced plate and arm and the CO adjusting lever

Bosch Systems

Throttle valve

The throttle valve housing is mounted directly on the intake manifold. It contains the main throttle valve (butterfly valve or throttle plate), the idle air passage and the connections to the auxiliary air valve and cold start valve.

Fuel distributor

The fuel distributor controls the amount of fuel passed into the injection system and distributes that fuel to the individual injectors. On the KE-Jetronic system fuel is delivered to the fuel distributor at a constant 79 psi, and the electrohydraulic actuator determines the flow rate into the lower chamber, regulating the amount of fuel flowing to the injectors. As the control plunger is moved in response to the air flow sensor, fuel passages are uncovered in the fuel distributor, allowing more or less fuel to pass through the distributor. The electrohydraulic actuator is also used to control the mixture (in response to signals from the ECU) for cold start and warm-up enrichment.

Since a fuel pressure drop would occur when the throttle is opened sharply, pressure valves within the distributor body maintain a constant fuel pressure at the metering passages. On later models (after 1977) a one-way valve is incorporated in the system to maintain fuel pressure after the engine is shut off. This is called the "push valve."

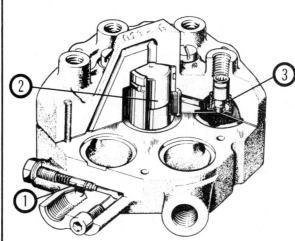

The fuel distributor is made up of a line pressure regulator (1) which controls the fuel pressure, a fuel control unit (2) which regulates and distributes the fuel to the injectors and pressure regulator valves for each cylinder (3)

Line pressure regulator

The line pressure regulator maintains a constant fuel pressure in the system while the engine is running, returning excess fuel to the tank to prevent overpressure. In addition, the regulator has a built-in rest pressure valve and a shut off valve, which function when the engine has been shut off to maintain a lower (approximately 35 psi) pressure in the system to prevent vapor lock and to keep the fuel in the system from draining back into the tank, making starting more difficult.

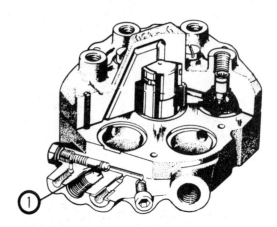

The line pressure regulator maintains a constant pressure in the fuel system while the engine is running, and the check valve (1) ensures that a lower, but still positive, pressure is maintained in the system when the engine is shut off to make restarting easier

K-Jetronic

Fuel accumulator

The fuel accumulator works in conjunction with the line pressure regulator to maintain the fuel delivery pressure in the fuel distributor at approximately 73 psi by opening a valve which returns excess fuel to the tank when the pressure rises above that level.

Fuel injectors

The fuel injectors for the K-Jetronic system are mechanical units, spring loaded and set to open at anything over a preset pressure, usually approximately 50 psi. The injectors spray fuel into the intake manifold upstream from the intake valves in a constant stream, so long as the fuel pressure remains high enough to hold the injectors open.

The injectors have no metering function. They are either open or closed, and the amount of fuel delivered by each injector is determined by the pressure maintained by the fuel distributor. The higher the pressure, the greater the fuel delivery.

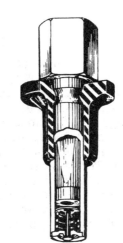

The K-Jetronic type of fuel injector contains a spring-loaded disc type valve which remains open as long as the engine is running and the fuel pressure is above a set level, usually approximately 50 psi. The amount of fuel injected is controlled by the fuel distributor, rather than by the opening and closing of the injector

Oxygen sensor

The oxygen sensor, located in the exhaust stream, is the center of the Lambda system, which enables the electronic control unit to control the mixture, and therefore the level of emissions.

When the mixture is at an ideal point, emissions are at a minimum, and the fuel/air ratio which is ideal for any given operating condition is referred to by Bosch as Lambda. The oxygen sensor constantly monitors the amount of unburned oxygen in the exhaust and the signal sent to the ECU is matched against the Lambda ideal. A signal from the ECU to the Fuel distributor richens or leans the mixture as necessary for proper emissions control.

The oxygen sensor only operates after it has been heated by the exhaust stream, so, to provide more rapid actuation of the Lambda system and thereby reduce the time when emissions are essentially uncontrolled, the oxygen sensor on some models contains an electrical heating element. This type of oxygen sensor can be identified by the use of three wires, rather than the two used on unheated sensors. The two types of sensors cannot be interchanged.

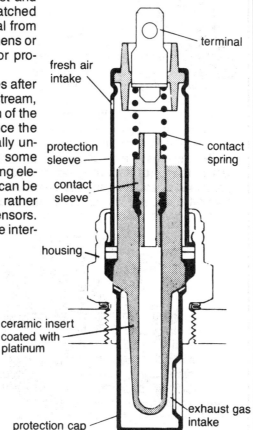

The oxygen sensor constantly samples the amount of oxygen remaining in the exhaust stream, and whenever the mixture is incorrect (too rich or too lean) the sensor sends a signal to the Electronic Control Unit to adjust the amount of fuel injected into the engine accordingly

Bosch Systems

Cold start valve

The cold start valve is an electronically controlled injector which supplies extra fuel when cold-engine operation requires a faster idle speed and richer mixture

1. Solenoid
2. Return spring
3. Actuator
4. Seal
5. Nozzle

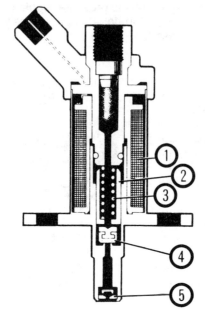

Under cold starting conditions extra fuel is required by the engine and it is the function of the cold start valve to supply that fuel. The valve is actually a fuel injector, activated by a solenoid, and the fuel is injected at a central location into the intake manifold to supplement the fuel being delivered by the individual injectors.

An engine temperature sensor and an electrically heated bimetallic strip are used to control the cold start valve, which is automatically shut off after approximately 8 seconds of operation to prevent flooding.

Thermo-time switch

The thermo-time switch is used to control the amount of time the cold start valve delivers extra fuel to the engine by sensing the temperature of the engine coolant and the length of time the starter is energized

The thermo-time switch regulates the amount of time the cold start valve is energized and prevents the opening of the cold start valve and subsequent mixture enrichment when the engine is warm.

The "thermo" section of the switch is a heater coil. When the starter is energized, current is passed through the thermo-time switch and into the cold start valve. As the current passes through the switch, the heater coil warms a bimetallic switch. It takes between 8 and 12 seconds for the heater coil to warm the bimetallic switch to the point where it opens the circuit, eliminating the extra fuel flow enrichment through the cold start valve.

The length of time the circuit is energized is controlled by the engine coolant temperature. The lower the coolant temperature, the longer it will take the heater coil to warm the bimetallic switch.

Warm-up regulator

The warm-up regulator also acts to enrich the mixture during cold engine operation by causing the fuel distributor to open the fuel metering ports an extra amount, delivering more fuel to the fuel injectors.

An electrically heated bimetallic spring works against the control pressure circuit diaphragm, increasing the fuel pressure in the fuel distributor by returning less excess fuel to the tank. As the bimetallic spring is heated, the pressure of the spring against the control pressure diaphragm decreases, until the engine is warmed and the warm-up regulator is removed from the circuit.

K-Jetronic

Auxiliary air regulator

The auxiliary air regulator acts as an air control system during cold engine operation and start-up. When the engine is cold, it bypasses extra air around the throttle plate, which is read as extra air flow by the ECU, which signals for additional fuel.

A bimetallic spring, activated by coolant temperature, and an electrical heater coil, to prevent excessive operation of the auxiliary air regulator, control a rotating disc inside the throttle chamber. The disc has a hole in it and when the hole lines up with a matching hole in the throttle chamber bypass, extra air is inducted into the engine. As the bimetallic spring heats up, it rotates the disc, gradually closing off the hole until all the air passing into the engine is directed through the main throttle chamber passage.

Fuel pump

The electric fuel pump maintains a constant flow of fuel from the tank to the fuel pressure regulator, running as long as the ignition is in the On position and the engine is running. A time delay circuit allows the pump to run during the starting procedure, even though the engine is not running. This circuit will automatically shut the fuel pump off after a certain amount of time if the ignition switch is On but the engine is not running.

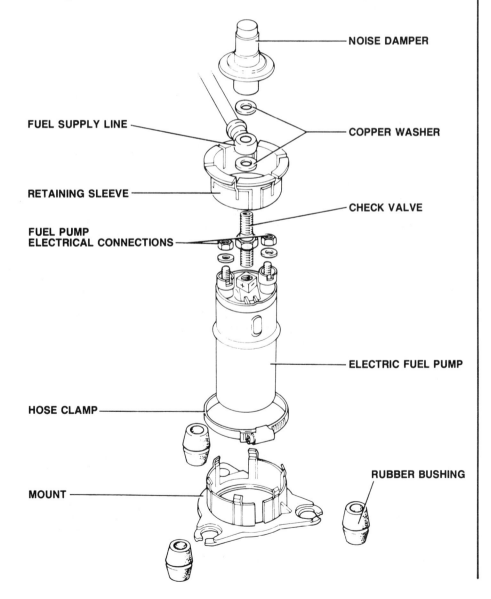

An electric fuel pump is utilized to provide the fuel distributor with fuel at sufficient pressure. The pump is both cooled and lubricated by the fuel passing through it, and it should never be run if the fuel tank is empty or if the fuel tank line is disconnected

Bosch Systems

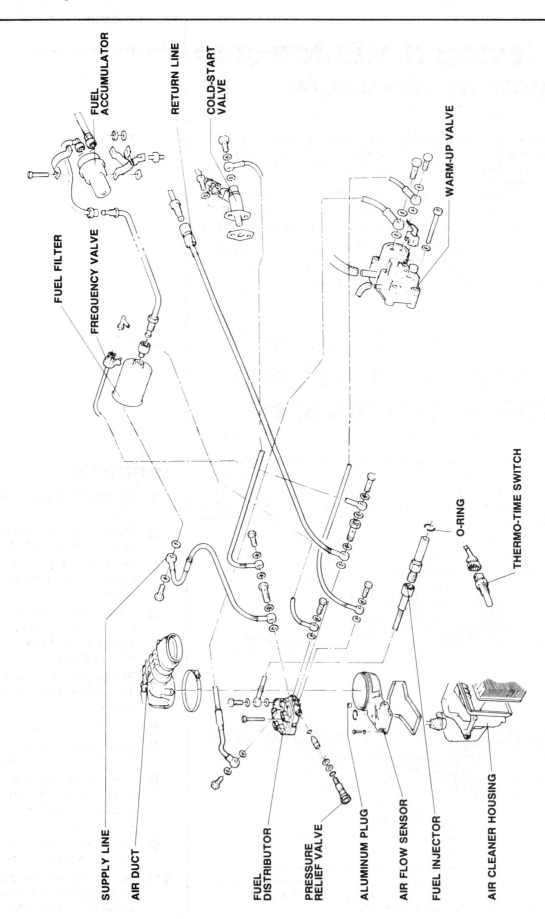

An exploded view of a typical K-Jetronic fuel injection system

K-Jetronic

7 Relieving K/KE-Jetronic fuel system pressure

The fuel system is kept under pressure even after the engine has been shut off (approximately 35 psi) to prevent vaporization of the fuel in the lines and to aid in starting. Before working on any part of the fuel system this pressure must be relieved.

To relieve the fuel pressure, disconnect the electrical connection to the cold start valve, then use a jumper cable to apply 12 volts from the battery directly to the electrical terminal of the cold start valve for ten seconds. Use extreme caution when making this connection as the connectors are very close together and a direct short is possible if care is not taken. This will eliminate the pressure in the system so long as the ignition switch is not turned to the On position. Reconnect the wire to the cold start valve.

8 Typical K/KE-Jetronic fuel system removal and installation procedures

Note: While installation on various makes and models of vehicles varies due to manufacturer installed component location, engine configuration (V-type, inline, flat opposed) and number of cylinders, the basic design of the K- and KE-Jetronic systems will be the same in all cases, and the following typical removal and installation procedure will, with only slight modifications, outline the procedure for your vehicle.

Clamps hold both the intake and output ducts to the air flow meter, and it is essential that a tight seal be maintained at this point to keep unmeasured air from entering the system. When removing the clamps and ducts check them carefully to make sure they are in good condition

Removal

1 Relieve the pressure in the fuel system.
2 Disconnect the negative cable from the battery.
3 Disconnect the electrical connector from the air flow meter.
4 Loosen the air intake and output duct clamps and separate the ducts from the air flow meter.
5 Remove the bolts securing the air flow meter to the bracket.
6 Remove the air flow meter.
7 Separate the throttle body from the air flow meter.
8 Remove the two screws and separate the throttle position switch from the throttle body.
9 Remove the fuel lines from the fuel pressure regulator.
10 Disconnect the fuel pressure regulator to intake manifold vacuum line.

Bosch Systems

11. Remove the fuel pressure regulator.
12. Remove the cold start valve electrical connector.
13. Disconnect the cold start valve fuel line.
14. Remove the two bolts and detach the cold start valve and gasket from the intake manifold.
15. Disconnect the electrical connector to the supplementary air regulator.
16. Disconnect the input and output air hoses from the supplementary air regulator.
17. Remove the bolts and detach the air regulator.
18. Disconnect the thermo-time switch electrical connector.
19. Drain the cooling system.
20. Remove the thermo-time switch.
21. Disconnect the coolant temperature sensor electrical connector.
22. Remove the coolant temperature sensor.
23. Disconnect the oxygen sensor electrical connector.
24. Remove the oxygen sensor from the exhaust manifold.
25. Remove the fuel injector-to-fuel rail clamp bolts.
26. Separate the fuel rail from the fuel injectors.
27. Remove the fuel injectors from the intake manifold.

Installation

28. Check the condition of the fuel injector O-rings, lubricate them with light oil, and install them in the intake manifold.
29. Press the fuel rail securely over the fuel injectors and install the fuel rail-to-injector clamps and bolts.
30. Apply anti-seize compound to the threads of the oxy-

Remove the electrical connector from the cold start valve

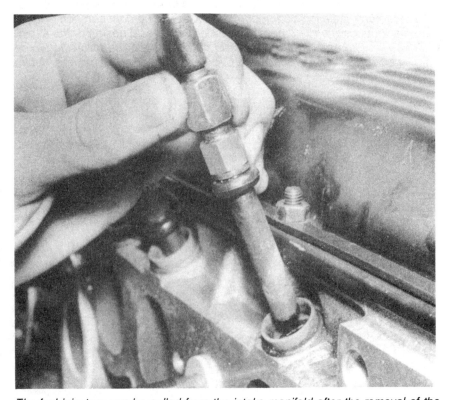

The fuel injectors can be pulled from the intake manifold after the removal of the fuel rail

K-Jetronic

After lubricating the O-rings with oil, use an open-end wrench to push the injectors firmly into the intake manifold

gen sensor and install the sensor in the exhaust manifold. Use caution not to get any of the anti-seize compound on the end of the sensor.

31 Connect the oxygen sensor electrical connector.

32 Wrap the threads with teflon tape, install the coolant temperature sensor and attach the electrical connector.

33 Install the thermo-time switch and attach the electrical connector.

34 Install the supplementary air regulator.

35 Install the input and output air hoses on the air regulator.

36 Install the electrical connector on the supplementary air regulator.

37 Using a new gasket, install the cold start valve in the intake manifold.

38 Install the cold start valve electrical connector.

39 Connect the fuel line to the cold start valve.

40 Install the fuel pressure regulator.

41 Connect the vacuum line between the fuel pressure regulator and the intake manifold.

42 Install the fuel lines on the fuel pressure regulator.

43 Install the throttle position sensor on the throttle body.

44 Connect the throttle body assembly to the air flow meter.

45 Install the air flow meter bracket bolts.

46 Attach the intake and output air ducts to the air flow meter.

47 Attach the electrical connector to the air flow meter.

48 Fill the cooling system.

49 Reattach the negative cable to the battery.

General Motors Fuel Injection Systems

Throttle Body System
Throttle body injection (TBI) 1
    General information
    Troubleshooting

Throttle body injection components 2
    Electronic control module
    Idle speed bypass valve
    Fuel pump
    Fuel pressure regulator

Relieving TBI fuel system pressure 3

Typical TBI system removal and installation procedures 4

Multi-port System
Multi-port fuel injection 5
    General information
    Troubleshooting

Port fuel injection components 6
    Electronic control module
    Air mass sensor
    Idle air control
    Fuel pump

Relieving port injection fuel system pressure 7

Typical port fuel injection system removal and installation procedures 8

Digital System
Digital fuel injection (DFI) 9
    General information

Digital fuel injection (DFI) components 10
    Throttle body
    Fuel injectors
    Fuel pressure regulator
    Throttle position switch
    Idle speed control
    Manifold absolute pressure sensor
    Barometric pressure sensor
    Electronic control unit
    Fuel pump

Sequential System
Sequential fuel injection (SFI) 11
    General information

Sequential fuel injection (SFI) components 12
    Electronic control module
    Air mass sensor
    Idle air control
    Fuel pump

Relieving SFI fuel system pressure 13

1 Throttle body injection (TBI)

General information

General Motors throttle body injection utilizes one or two fuel injection nozzles, mounted in a throttle body on the intake manifold, rather than individual injectors for each cylinder, mounted in or near the intake port. Because of this, the system is somewhat less efficient than port-type injection, but it is also simpler and more economical to produce. In effect, the throttle body injection system is very close to a conventional carburetor in operation, but without the necessity for many of the carburetor's complex systems.

The injector is a standard electrically-operated Bosch-type solenoid. Fuel pressure in the system is maintained at 10 psi, and the Electronic Control Module (ECM) controls a pintle valve in the injector, the opening time controlling the amount of fuel delivered to the engine.

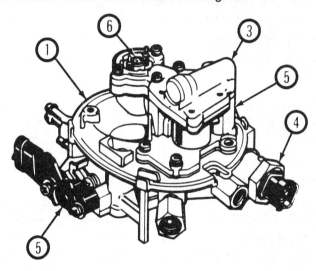

The General Motors throttle body is a compact unit which encloses all of the essential fuel injection components in a unit the same size as a conventional carburetor
1. *Throttle body*
2. *Fuel body*
3. *Fuel meter cover and pressure regulator*
4. *Idle air control valve*
5. *Throttle position sensor*
6. *Fuel injector*

Troubleshooting

Note: *All diagnostic causes listed below relate only to the fuel injection system. Quite often the symptom described may have causes other than in the fuel injection system. As an example, "Pinging sound under load" may well be due to a fault in the cooling system or the ignition timing, although such causes are not listed here.*

Pinging sound under load:
Faulty MAP sensor
Transmission converter clutch switch malfunctioning

Uneven idle when cold:
Air cleaner damper door not operating properly
Intake system vacuum leak
Sticking throttle linkage
Idle air control motor malfunction

Uneven idle when hot:
Air cleaner damper door not operating properly
Intake system vacuum leak
Idle air control motor malfunction

Engine stalls when cold:
Air cleaner damper door not operating properly
Intake system vacuum leak
Idle air control motor malfunction
Restricted fuel line or filter
System fuel pressure incorrect

General Motors

Troubleshooting (continued)

Engine stalls when hot:
 Air cleaner damper door not operating properly
 Intake system vacuum leak
 Idle air control motor malfunction
 Restricted fuel line or filter
 System fuel pressure incorrect

Engine hard to start:
 Restricted fuel line or filter
 System fuel pressure incorrect
 Fuel pump relay malfunction
 Throttle position sensor sticking
 Leaking fuel line check valve
 Injector leaking

Engine misses:
 Intake system vacuum leak
 Air cleaner damper door not operating properly
 Restricted fuel line or filter
 System fuel pressure incorrect
 MAP system air leak
 Throttle position sensor sticking
 Injector leaking
 Injector plugged

Engine hesitates:
 Intake system vacuum leak
 Air cleaner damper door not operating properly
 System fuel pressure incorrect
 MAP system air leak
 Throttle position sensor sticking
 Defective injector

Engine surges:
 Air cleaner damper door not operating properly
 Intake system vacuum leak
 Fuel filter excessively dirty
 System fuel pressure incorrect

Engine runs after switching off (diesels):
 Sticking or leaking injector

Loss of power:
 Dirty air filter
 Air cleaner damper door not operating properly
 MAP sensor malfunctioning
 MAP system air leak
 System fuel pressure incorrect
 Throttle position sensor sticking
 Defective injector

Excessive fuel consumption:
 Air cleaner damper door not operating properly
 Intake system vacuum leak

Throttle Body (TBI)

2 Throttle body injection components

In determining how much fuel is required at any given moment, the Electronic Control Module processes signals from sensors recording engine temperature, throttle position, vehicle speed, manifold vacuum and the exhaust oxygen content, updating the information every 1/10 second.

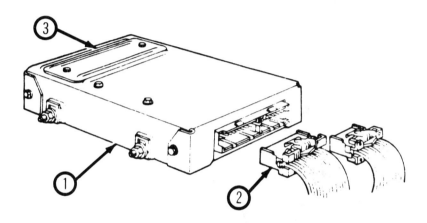

Electronic control module

The Electronic Control Module (1) receives signals from a variety of sensors through the ECM harness (2), and the Programmable Read Only Memory (3) interprets these signals and directs the operation of the fuel injection unit

The idle speed on the TBI is also controlled by the ECM, through an idle air control valve. When the engine is cold or when accessory loads, such as air conditioning, necessitate a higher idle speed, the ECM, through an idle air control motor, opens the IAC valve to provide more air into the system, and at the same time the ECM increases the fuel injector open time to maintain the proper fuel/air ratio.

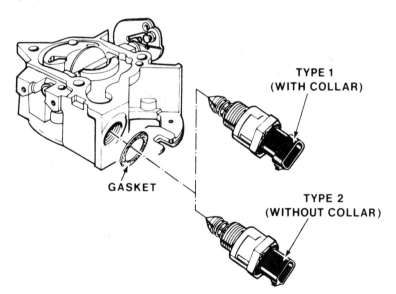

Idle air control valve

The idle air control assembly (two types are used) is opened by the Electronic Control Module when the engine is cold to allow more air into the system. This, combined with additional fuel from the injector, increases the idle speed

65

General Motors

Fuel pump

While the open time of the injector is controlled by the ECM, the actual fuel delivery is determined by the fuel pressure, which must be maintained at a constant level. This is accomplished through the use of a two-stage turbine pump, driven by an electric motor and mounted in the fuel tank. The pump intake is mounted in a special sump in the fuel tank so that a low fuel level, combined with hard cornering, acceleration or deceleration, cannot uncover the pump inlet, causing a pressure drop.

Fuel pressure regulator

A fuel pressure regulator is designed into the throttle body, which maintains the required 10 psi pressure in the system by returning unneeded fuel to the tank. On the systems using two injector nozzles, a fuel pressure compensator is used to maintain even pressure at the second injector even though the first is open, bleeding off pressure in the line.

The fuel pressure regulator maintains a constant pressure in the fuel system. The regulator cover, held on by four screws, should never be removed from the fuel metering assembly

3 Relieving TBI fuel system pressure

The fuel system remains under pressure even after the engine has been shut off for an extended time. Therefore it is necessary to relieve the pressure in the fuel system before any work is done on fuel injection components or lines.

To relieve the fuel system pressure, remove the fuel pump fuse from the fuse block, disabling the pump. Start the engine, letting it run until it dies for lack of fuel, then crank the engine over for several seconds with the starter to insure that all pressure has been eliminated from the system. Be sure to replace the fuel pump fuse when work is completed on the fuel system.

Throttle Body (TBI)

4 Typical TBI system removal and installation procedures

Note: While installation on various makes and models of vehicles varies due to manufacturer installed component location, engine configuration (V-type or inline) and number of cylinders, the basic design of the throttle body injection system will be the same in all cases, and the following typical removal and installation procedure will, with only slight modifications, outline the procedure for your vehicle.

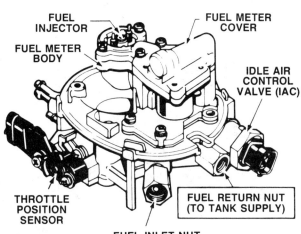

Remove the fuel inlet and return lines, then disconnect the wires to the idle air control valve, throttle position sensor and fuel injector

Removal

1. Relieve the pressure in the fuel system.
2. Remove the negative cable from the battery.
3. Remove the air cleaner.
4. Disconnect the wires leading to the idle air control, throttle position sensor and fuel injector.
5. Disconnect the throttle linkage and return spring from the throttle body housing.
6. If so equipped, disconnect the cruise control linkage from the throttle body housing.
7. Use tape to label the installed locations of all vacuum lines leading to the throttle body housing, then disconnect them.
8. Disconnect the fuel lines from the throttle body housing.
9. Remove the three mounting bolts and lift the throttle body off, along with the gasket.
10. Some throttle position sensors are adjusted at the factory and spot welded in place to retain their settings. Others are held in place with screws.

General Motors

Removal
(continued)

11 If the throttle position sensor must be replaced, obtain a sensor service kit from a General Motors dealer.

12 Using a 5/16 inch drill bit, drill completely through the two throttle position sensor screws in the access holes in the throttle body base to remove the spot welds holding the screws in place.

13 Remove the throttle position sensor screws, lockwashers and retainers and lift off the throttle position sensor. Discard the screws.

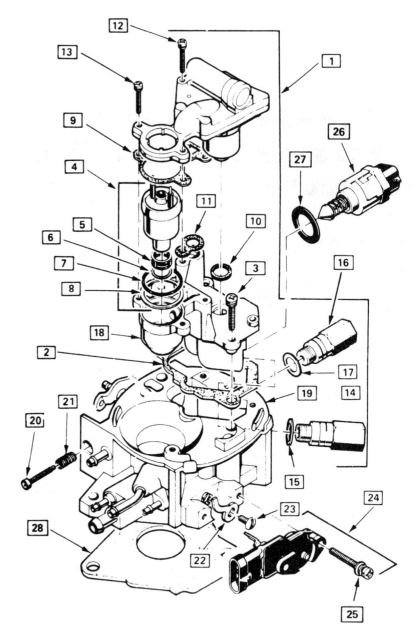

Exploded view of the throttle body assembly

1 Fuel metering assembly
2 Fuel metering body gasket
3 Fuel metering body screws
4 Fuel injector assembly
5 Fuel injector filter
6 Small fuel injector O-ring
7 Large fuel injector O-ring
8 Fuel injector back-up washer
9 Fuel metering cover-to-injector gasket
10 Pressure regulator dust seal
11 Fuel metering body outlet gasket
12 Fuel metering body screws
13 Fuel metering body screws
14 Fuel inlet nut
15 Fuel inlet nut gasket
16 Fuel outlet nut
17 Fuel outlet nut gasket
18 Fuel metering body assembly
19 Throttle body assembly
20 Idle stop screw
21 Idle stop screw spring
22 Throttle position sensor lever
23 Throttle position sensor lever screw
24 Throttle position sensor assembly
25 Throttle position sensor screws
26 Idle air control assembly
27 Idle air control assembly gasket
28 Flange mounting gasket

Throttle Body (TBI)

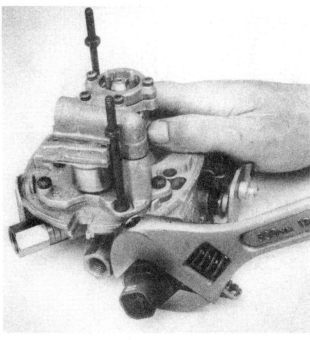

Remove the idle air control valve from the throttle body

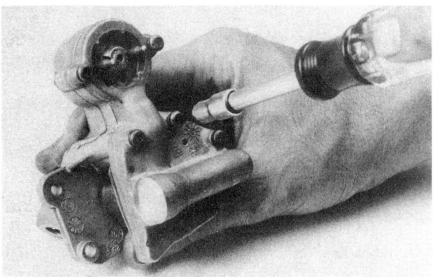

Remove the screws holding the fuel metering body to the throttle body and remove the metering unit

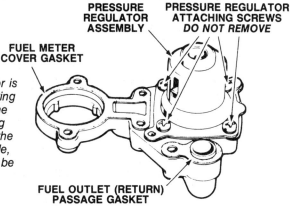

The fuel pressure regulator is attached to the fuel metering body with four screws. The regulator contains a spring under high pressure and the assembly is not rebuildable, so the screws should not be removed at any time

14 Unscrew the idle air control assembly from the throttle body.

15 The fuel meter cover assembly, including the fuel pressure regulator, is serviced only as a complete unit. If either the regulator or cover requires replacement, the entire assembly must be replaced.
Warning: The fuel pressure regulator is enclosed in the fuel meter cover. It is under heavy spring tension and the four screws securing the pressure regulator to the fuel meter cover should not be removed, as personal injury could result.

16 Remove the fuel meter cover mounting screws and separate the cover from the body. Again, these are the screws holding the cover to the body, not the four screws holding the regulator to the cover. Note the positions of the two shorter screws, because they must be reinstalled in the same positions.

General Motors

Removal (continued)

17 Place a screwdriver shank on top of the fuel meter cover gasket surface to serve as a fulcrum and, using another screwdriver, carefully pry the fuel injector out of the fuel meter body. A gasket can be used on the fuel meter cover to prevent the surface from being scratched by the screwdriver during the procedure.

18 Do not push the injector out from underneath as this could damage the injector tip.

19 Clean all metal parts in a cold immersion-type cleaner and blow them dry with compressed air. *Caution: The throttle position sensor, idle air control assembly, fuel meter cover/fuel pressure regulator, fuel injector, fuel injector filter and all diaphragms and other rubber and plastic parts should not be immersed in cleaner, as it will cause swelling, hardening and distortion, and in some cases might even dissolve the part. Clean these parts by hand with clean shop rags. Make sure all air and fuel passages are clear. Any thread locking compound on the idle air control assembly mounting threads should be allowed to remain.*

Two screwdrivers can be used to lever the fuel injector from the recess in the throttle body

Disassembled view of the throttle body unit

Throttle Body (TBI)

Using new gaskets, install the fuel inlet and return nuts

4.14 Check the condition of the small O-ring and the fuel filter on the base of the fuel injector

Installation

20 Attach the fuel meter body to the throttle body.

21 Apply thread locking compound, supplied in the service kit, to the threads of the fuel meter body mounting screws. If thread locking compound is not provided, Threadlock Adhesive 262 or the equivalent may be substituted. Do not use a compound of higher strength than recommended, as it may promote screwhead breakage or prevent removal of the screws when service is again required.

22 Install the fuel meter body mounting screws and lockwashers.

23 Install the fuel feed and return nuts, with new gaskets, in the fuel meter body.

24 Check and clean the fuel filter on the base of the injector. To remove the filter, carefully rotate it back-and-forth to pull it off. The filter is installed by pushing it into the injector until it is seated.

General Motors

Installation
(continued)

25 Whenever the injector is removed, new O-rings should be installed. Remove the large O-ring and steel backup washer from the injector cavity, then remove the small O-ring from the bottom of the injector.

26 Lubricate a new small O-ring with lithium-based grease and push it over the nozzle end of the injector so it is seated against the fuel filter.

27 Install the steel backup ring washer in the meter body injector cavity.

28 Lubricate the large O-ring with lithium-based grease and install it in the injector cavity, directly above the washer. When properly installed, the O-ring is flush with the fuel meter body surface.

29 *Caution: Do not attempt to bypass this procedure by installing the backup washer and O-ring after the injector is located in the cavity, as this would prevent proper seating of the O-ring.*

30 Use a twisting motion to install the injector into the cavity. Be sure it is fully seated and that the raised lug in the injector base is aligned with the notch in the fuel meter body.

31 Install the fuel meter cover.

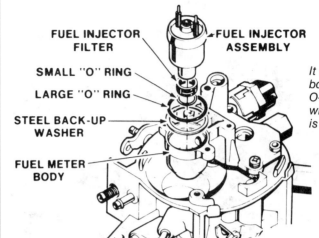

It is recommended that both the large and small O-rings be replaced whenever the fuel injector is removed

Lubricate the small O-ring with lithium-based grease and insert the injector into the body

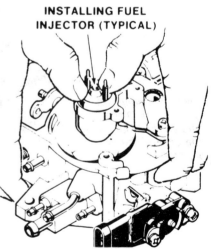

Firmly press the injector into the cavity, making sure the raised lug is aligned with the notch in the body

Throttle Body (TBI)

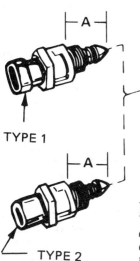

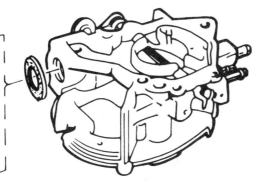

Before installing the idle air control valve the distance between the end of the housing and the tip of the valve must be measured

Two types of idle air control valves are used, Type 1 with a collar and Type 2 without a collar, and the method for setting the valve extension distance (A) is different for the two types

32 Before installing the idle air control, measure the distance between the end of the idle air control assembly housing and the tip of the conical valve. The valve should be extended from the housing no more than 1.260-inch (32 mm) on 1982 models or 1.125-inch (28.6 mm) on later models or damage may occur to the motor when it is installed.

33 If the distance is too large, first determine if the assembly is a Type 1, with a collar at the end of the idle air control assembly, or a Type 2, without a collar.

34 The pintle on a Type 1 can be retracted by pushing on the end of the pintle until it is retracted sufficiently.

35 On a Type 2, push the pintle in and attempt to turn it clockwise. If it turns, continue turning until it is properly set.

36 If the pintle will not turn, exert firm hand pressure to retract it. *Note: If the pintle is turned, be sure the spring returns to the original position with the straight portion of the spring end aligned with the flat surface under the pintle head.*

37 Install a new gasket on the idle air control unit and screw it into the throttle body housing.

38 With the throttle plate in the closed (idle) position in the throttle body, install the throttle position sensor onto the throttle body with the retainers and the two new screws and lockwashers which came in the service kit. Do not reuse the old screws.

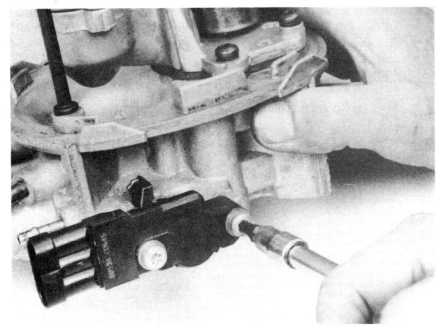

When installing a new throttle position sensor be sure to use the new screws supplied with the kit

General Motors

Installation (continued)

39 Be sure the throttle position sensor pickup lever is located above the tang on the throttle actuator lever. It is recommended that a thread-locking agent be used on the throttle position sensor screw threads.

40 Clean all old gasket material from the intake manifold mating surface, then install a new gasket.

41 Place the throttle body on the intake manifold and install the nuts.

42 Connect the fuel lines to the throttle body assembly. Be sure the fuel line O-rings are not nicked, cut or damaged before connecting the lines.

43 Install three jumper wires between the throttle position sensor connectors and the harness connector.

44 Connect the negative battery cable to the battery and turn the ignition to On.

45 Connect a voltmeter to terminals A and B.

46 Turn the throttle position sensor to the point where the voltmeter reads .54 volts $\pm$ 0.075 volts and tighten the screws.

47 Remove the jumper wires and connect the harness connector to the throttle position sensor.

48 Turn the ignition switch to Off.

49 If equipped, install the cruise control linkage.

50 Install the throttle linkage and return spring.

51 Connect the electrical connectors to the idle air control unit, throttle position sensor and fuel injector.

52 Install the air cleaner.

53 Start the engine and check for fuel leaks.

54 Allow the engine to reach normal operating temperature.

55 On manual transmission equipped vehicles the idle speed will automatically be controlled when normal operating temperature is reached.

56 On vehicles with an automatic transmission, the idle air control unit will begin controlling idle speed when the engine is at normal operating temperature and the transmission is shifted into Drive.

57 If the idle speed is too high and does not regulate back to normal after a few moments, operate the vehicle at a speed of 45 mph (72 kph). At that speed the electronic control module will command the idle air control pintle to extend fully to the mating seat in the throttle body, which will allow the electronic control module to establish an accurate reference point with respect to pintle position. Proper idle regulation will result.

Multi-port

5 Multi-port fuel injection

General information

The General Motors multi-port fuel injection system is used on certain models produced by Buick, Cadillac, Chevrolet, Oldsmobile and Pontiac. The system utilizes an injector solenoid at each intake port.

The injector is a standard Bosch-type electrically-operated solenoid. Fuel pressure in the system is maintained at 28-to-36 psi and the Electronic Control Module (ECM) controls the injector, the opening time regulating the amount of fuel delivered to the engine.

All of the injectors operate simultaneously, and half of the fuel required by each cylinder is injected with each injector pulse, the pulses coming once with each crankshaft revolution. Therefore, every other revolution, when the intake valve opens, the full charge needed is available in the intake port for ingestion into the cylinder.

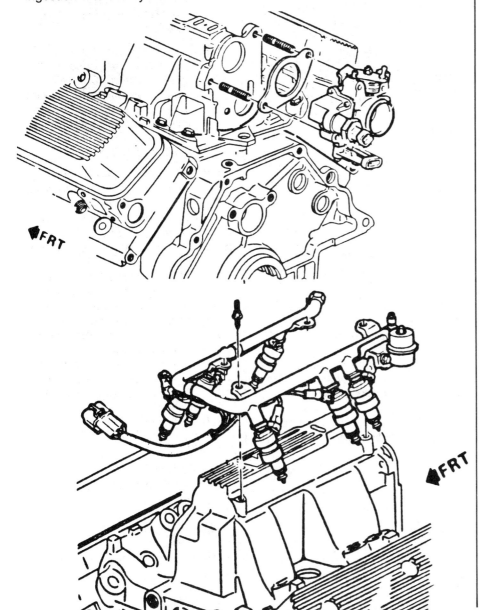

The General Motors port fuel injection system utilizes a throttle body assembly mounted on the intake manifold and an electrically-operated solenoid-type fuel injector in each intake port

General Motors

Troubleshooting

Note: All diagnostic causes listed below relate only to the fuel injection system. Quite often the symptom described may have causes other than in the fuel injection system. As an example, "Engine will not start" may well be due to a fault in the ignition system, although such causes are not listed here.

Hard to start:
Incorrect fuel pressure
Malfunctioning cold start valve
Throttle position sensor sticking
High TPS voltage with the throttle closed
Coolant sensor malfunction
Fuelpump check valve defective
Injector fuses blown
Mass air flow sensor malfunction

Hesitation:
Incorrect fuel pressure
Air leak between mass air flow sensor and throttle body
Throttle position sensor sticking
Alternator output voltage incorrect
Mass air flow sensor burn-off circuit malfunction

Surge:
Air leak between mass air flow sensor and throttle body
Alternator output voltage incorrect
Fuel filter dirty
Fuel pressure incorrect
Oxygen sensor contaminated

Loss of power:
Air filter dirty
Air leak between mass air flow sensor and throttle body
Fuel filter dirty
Fuel pressure incorrect
Alternator output voltage incorrect

Pinging sound under acceleration:
Fuel pressure incorrect

Missing:
Injector malfunction
Faulty electronic control module
Restricted fuel filter
Fuel pressure incorrect

Backfire:
Loose mass air flow sensor electrical connector
Loose mass air flow sensor air duct

Rough idle:
Sticking throttle linkage
Malfunctioning idle air control system
Alternator output voltage incorrect
Malfunctioning mass air flow sensor
Contaminated oxygen sensor
Faulty fuel pressure regulator

Multi-port

6 Port fuel injection components

Electronic control module

In determining how much fuel is required at any given moment, the Electronic Control Module processes signals from sensors recording engine coolant temperature, exhaust oxygen content, throttle position, intake air mass, engine rpm, vehicle speed and accessory load.

Air mass sensor

The intake air mass sensor measures the mass (weight) of the air entering the intake manifold, rather than the volume, which gives better control of the air/fuel mixture. The mass of any given volume of air is dependent upon the temperature of the air. Cold air weighs more (has more mass) than warm or hot air, and therefore the colder the air is, the more fuel is required to maintain the proper fuel/air ratio.

On the Bosch air mass sensor system a platinum wire is used to measure the intake air mass. On the General Motors system, however, a heated film is used. The temperature of the incoming air is constantly monitored and the air mass sensor plate is maintained at exactly 75 degrees above the temperature of the air. Since the incoming air, passing over the plate, acts to cool the plate, the measurement of the amount of electrical energy needed to maintain that 75 degree temperature differential accurately indicates the mass of air entering the engine, and therefore the amount of fuel necessary to mix with that mass to provide the proper fuel/air ratio.

Idle air control

The idle speed on the MFI system is also controlled by the electronic control module, through an idle air control bypass channel. When the engine is cold or when accessory loads, such as air conditioning, necessitate a higher idle speed, the ECM, through a stepper motor, opens the bypass pintle valve to provide more air into the system. At the same time, the ECM increases the fuel injector open time to maintain the proper fuel/air ratio.

Fuel pump

While the open time of the injectors is controlled by the electronic control module, the actual fuel delivery is determined by the fuel pressure, which must be maintained at a constant level. This is done by a positive displacement roller vane pump mounted in the fuel tank. The pump intake is mounted in a special sump in the fuel tank so that a low fuel level, combined with hard cornering, acceleration or deceleration, cannot uncover the pump inlet, causing a pressure drop.

A fuel pressure regulator maintains the required pressure in the system by returning unneeded fuel to the tank. A fuel accumulator is used to dampen the pulses which would otherwise be set up in the fuel rail and fuel line when the injectors open and close.

General Motors

7 Relieving port injection fuel system pressure

Before disconnecting any components of the fuel system the pressure in the system must be relieved. Note that this pressure can remain in the system long after the engine has been shut off.

To relieve the system pressure, first remove the fuel pump fuse from the fuse block, disabling the fuel pump. Start the engine and let it run until it dies from lack of fuel, then turn the engine over several times with the starter to eliminate all pressure in the system. Be sure to replace the fuel pump fuse after completing work on the system.

8 Typical port fuel injection system removal and installation procedures

Removal

1. Disconnect the negative cable from the battery.
2. Release the locking clamp holding the front air cleaner duct to the mass air flow sensor.
3. Remove the clamps from either end of the flexible air duct between the mass air flow sensor and the throttle body, then detach the duct.
4. Disconnect the electrical connector from the mass air flow sensor.

Note: While installation on various makes and models of vehicles varies due to manufacturer installed component location, engine configuration (V-type or inline) and number of cylinders, the basic design of the General Motors multi-point fuel injection system will be the same in all cases, and the following typical removal and installation procedure will, with only slight modifications, outline the procedure for your vehicle.

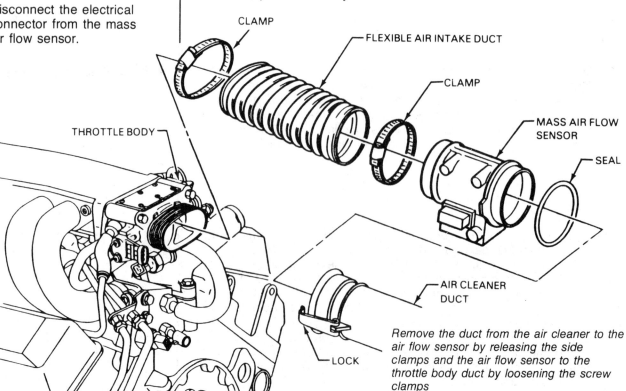

Remove the duct from the air cleaner to the air flow sensor by releasing the side clamps and the air flow sensor to the throttle body duct by loosening the screw clamps

Multi-port

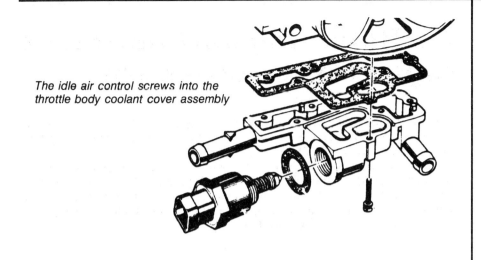

The idle air control screws into the throttle body coolant cover assembly

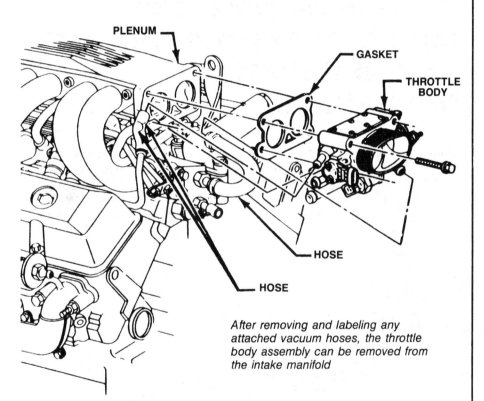

After removing and labeling any attached vacuum hoses, the throttle body assembly can be removed from the intake manifold

5 Disconnect the mass air flow sensor bracket and remove the mass air flow sensor.

6 Locate the oxygen sensor in the exhaust manifold.

7 Disconnect the electrical connector.

8 Remove the oxygen sensor.

9 Disconnect the electrical connector to the throttle position sensor on the throttle body assembly.

10 Remove the throttle position sensor screws and lift off the sensor.

11 Disconnect the electrical connector to the idle air control valve.

12 Remove the idle air control valve from the throttle body assembly.

13 Disconnect and label any vacuum lines attached to the throttle body.

14 Disconnect the two coolant hoses.

15 Remove the throttle cable.

16 If equipped with an automatic transmission, remove the throttle valve cable.

17 If equipped, remove the cruise control cable.

18 Remove the throttle body retaining bolts and the trottle body.

19 Disconnect the thermal-time switch electrical connector.

20 Remove the thermal-time switch.

21 Remove the idle air control valve coolant cover screws and the cover assembly.

22 Remove the plenum chamber.

General Motors

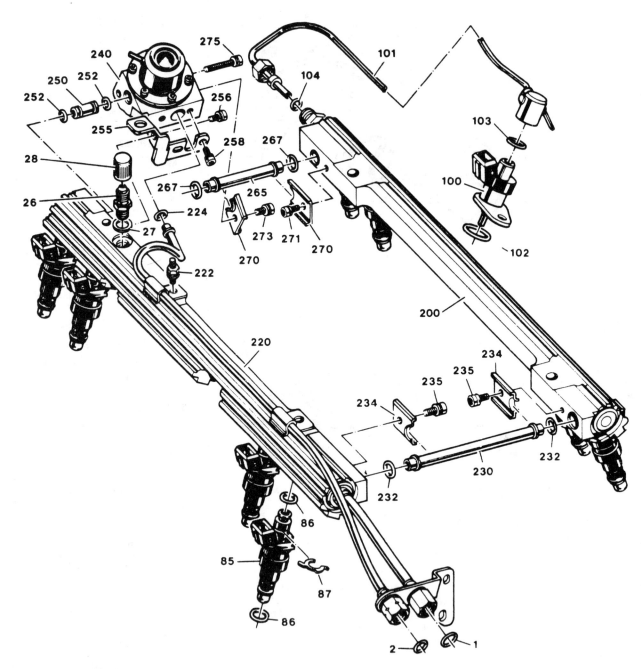

Exploded view of the fuel pressure regulator, cold start valve, fuel rails and fuel injectors

- 1 Fuel inlet O-ring
- 2 Fuel return O-ring
- 26 Fuel pressure connection assembly
- 27 Fuel pressure connection seal
- 28 Fuel pressure connection cap
- 85 Fuel injector
- 86 Fuel injector O-ring
- 87 Fuel injector retainer clip
- 100 Cold start valve
- 101 Tube and body assembly
- 102 Cold start valve O-ring
- 103 Cold start valve body O-ring
- 104 Tube O-ring
- 200 Left fuel rail and plug assembly
- 222 Right fuel rail and plug assembly
- 224 Fuel outlet tube O-ring
- 230 Front crossover tube
- 232 Front crossover tube O-ring
- 234 Crossover tube retainer
- 235 Retainer attaching screw assembly
- 240 Fuel pressure regulator and base assembly
- 250 Base to rail connector
- 252 Connector O-ring
- 255 Fuel pressure regulator bracket
- 256 Bracket to rail screw assembly
- 258 Bracket to base screw assembly
- 265 Rear crossover tube
- 267 Crossover tube O-ring
- 270 Rear crossover tube retainer
- 271 Retainer to left rail screw assembly
- 273 Retainer to base screw assembly
- 275 Base to right rail screw assembly

Multi-port

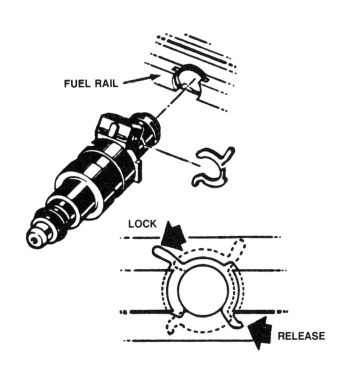

To remove the injectors from the fuel rail the retaining clips must be turned to the release position

Removal
(continued)

23 Remove the fuel lines where they connect to the fuel rail assembly.

24 Disconnect the electrical connector from the cold start valve.

25 Remove the cold start valve tube.

26 Remove the cold start valve retaining bolt and separate the cold start valve from the fuel rail. Bend back the tab to allow the valve to be unscrewed from the valve body.

27 Remove the fuel injector electrical connectors.

28 Remove the fuel rail retaining bolts and detach the fuel rails and injectors.

29 Turn the fuel injector retaining clips to the unlocked position.

30 Remove the fuel injectors from the fuel rails.

31 Remove the front and rear crossover tubes between the fuel rails.

32 Separate the left and right fuel rails.

33 Remove the bracket screws, rotate the pressure regulator and separate the pressure regulator from the fuel rails.

Installation

34 Replace the O-rings on the fuel pressure regulator.

35 Lubricate the new O-rings with light oil.

36 Install the pressure regulator brackets, carefully fit the fuel lines into position and rotate the pressure regulator into position.

37 Install new O-rings on the front and rear crossover tubes and reunite the left and right fuel rails.

General Motors

Installation
(continued)

38. Inspect the fuel injector O-rings and replace any which are cracked, broken or hardened.

39. Lubricate the upper O-rings with light oil.

40. Insert the fuel injectors into the fuel rails and turn the retaining clips to the locked position.

41. Coat the lower injector O-rings with light oil and insert the injectors into the intake manifold.

42. Install the fuel rail retaining bolts.

43. Install the fuel injector electrical connectors.

44. Install the fuel lines.

45. Install new O-rings on the cold start valve.

46. Install a new O-ring on the cold start valve tube.

47. Install the cold start valve and cold start valve tube into the fuel rail.

48. Turn the cold start valve fully into the valve body.

49. Back the valve out one full turn, until the electrical connector is in the up position.

50. Bend the tab up to limit the rotation of the valve to less than one turn.

51. Connect the electrical connector to the cold start valve.

52. Install a new plenum gasket.

53. Install the plenum chamber.

54. Install a new idle air control/coolant cover gasket on the throttle body.

55. Install the idle air control/coolant cover.

56. Coat the threads of the thermo-time switch with sealant and install the switch.

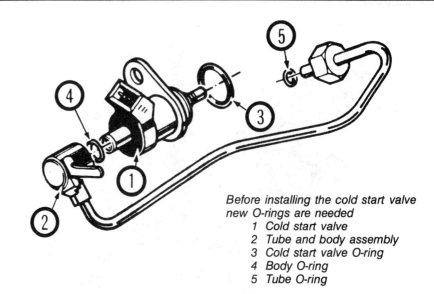

Before installing the cold start valve new O-rings are needed
1. Cold start valve
2. Tube and body assembly
3. Cold start valve O-ring
4. Body O-ring
5. Tube O-ring

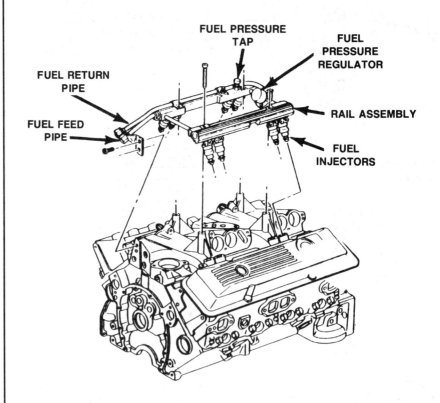

Drop the fuel rail assembly down over the intake manifold, making sure the injectors seat properly

Multi-port

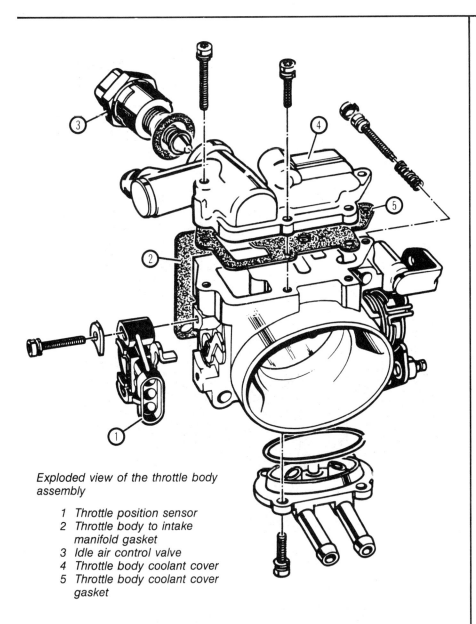

Exploded view of the throttle body assembly

1. Throttle position sensor
2. Throttle body to intake manifold gasket
3. Idle air control valve
4. Throttle body coolant cover
5. Throttle body coolant cover gasket

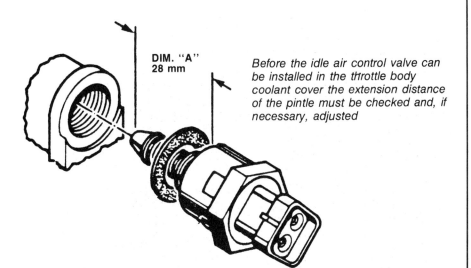

Before the idle air control valve can be installed in the throttle body coolant cover the extension distance of the pintle must be checked and, if necessary, adjusted

57 Connect the thermo-time switch electrical connector.

58 Install the throttle body onto the plenum chamber.

59 If used, install the cruise control cable.

60 If equipped with an automatic transmission, install the throttle valve cable or linkage.

61 Install the throttle cable.

62 Attach any disconnected vacuum lines to the throttle body.

63 Measure the distance from the gasket surface of the idle air control valve to the tip of the pintle.

64 If the distance is greater than 1-1/8 inches it must be reduced to prevent damage on installation.

65 If the idle air control valve has a collar around the electrical connector end, push the pintle in firmly to retract it.

66 If the idle air control valve does not have a collar, compress the pintle retaining spring towards the body and try to turn the pintle clockwise.

67 If the pintle will turn, continue turning it clockwise until it extends less than 1-1/8 inches. Return the spring to its original position, with the straight part of the spring end lined up with the flat surface under the pintle head.

68 If the pintle will not turn, use firm hand pressure to retract it.

69 Install a new gasket on the idle air control valve and install the valve in the idle air control/coolant cover assembly.

General Motors

70 With the throttle plate closed (idle position), install the throttle position sensor on the throttle body assembly, making sure the pickup lever is above the tang on the throttle actuator lever.

71 Install three jumper wires between the throttle position sensor connectors and the harness connector.

72 Connect a voltmeter to terminals A and B.

73 Connect the battery negative cable and turn the ignition switch to On.

74 Rotate the throttle position sensor to the position where the voltmeter reads .54 volts ± 0.075 volts and tighten the screws.

75 Turn the ignition switch to Off.

76 Remove the jumper wires and connect the harness connector to the throttle position sensor.

77 If reinstalling the old oxygen sensor, coat the threads with anti-seize compound. New sensors will come with the threads already coated.

78 Install the oxygen sensor and connect the electrical connector.

79 Put the mass air flow sensor in place and connect the attaching bracket.

80 Connect the electrical connector to the mass air flow sensor.

81 Install the flexible air duct between the mass air flow sensor and the throttle body and securely tighten the clamps at each end of the duct.

82 Make sure the sealing ring is in place on the mass air flow sensor, then attach the front air cleaner duct and close the locking clamp.

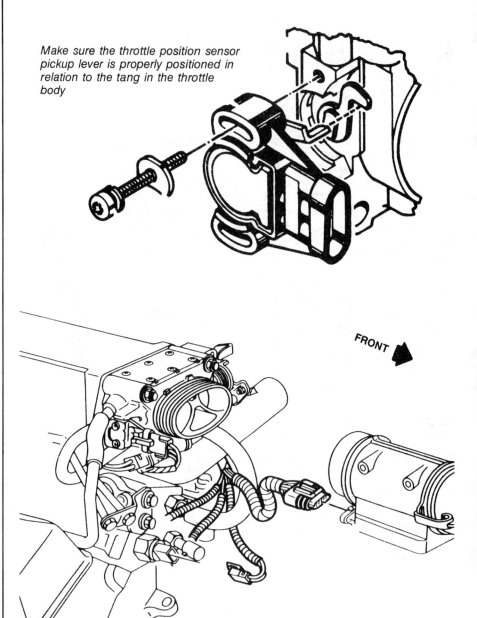

Make sure the throttle position sensor pickup lever is properly positioned in relation to the tang in the throttle body

Make sure the coolant temperature sensor and mass air flow sensor connectors are firmly installed

9 Digital fuel injection (DFI)

General information

Some Buick, Oldsmobile and Cadillac models are equipped with Digital Electronic Fuel Injection, which is a throttle body type system with a pair of fuel injectors located above the throttle plate. The injectors are activated, in alternating sequence, by signals from the Electronic Control Module (ECU).

An electric fuel pump supplies pressure to the fuel system, with a pressure regulator maintaining a steady pressure differential between the fuel system side of the injector and the intake manifold pressure. With a constant pressure differential between these two points, the amount of fuel injected into the engine is determined by the amount of time the injector is open.

The electronic control unit, in determining how long to hold the injector solenoids open, receives signals from the Manifold Absolute Pressure sensor (MAP), Manifold Air Temperature sensor (MAT), Throttle Position Sensor (TPS), barometric pressure sensor (BARO), Coolant Temperature Sensor (CTS) and the ignition coil, which provides information on engine rpm. As a result of all of these signal inputs, the ECU is able to determine the amount of air flowing into the engine, the demands being made on the engine, and therefore the amount of fuel required at any given moment.

10 Digital fuel injection (DFI) components

Throttle body

The throttle body is mounted on the intake manifold and is simply a throttle bore casting with the throttle plates controlling the amount of air which enters the induction system. Mounted on or in the throttle body, however, are the components which make up the heart of the digital fuel injection system.

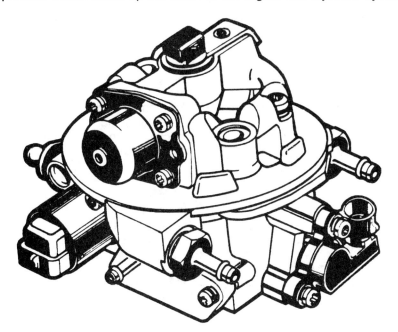

The digital fuel injection system utilizes a throttle body unit mounted on the intake manifold

General Motors

Fuel injectors

The two fuel injectors, mounted in the throttle body just above the throttle plate, are actuated by an internal electrical solenoid, which lifts a ball off a seat to allow the fuel to pass through the injector. The ball has only two positions — fully open and fully closed — and the amount of fuel injected is controlled by the length of time the injector ball is held in the open position.

Fuel pressure regulator

The fuel pressure regulator, which is mounted inside the throttle body, controls the amount of fuel allowed to flow through the injectors by maintaining a constant 10.5 psi pressure differential between the fuel side of the injectors and the manifold pressure. To control the fuel pressure excess fuel is returned to the fuel tank.

Throttle position switch

The throttle position sensor is mounted on the throttle body, from where it sends a signal to the Electronic Control Module when the throttle is in either wide-open or idle position

The throttle position switch sends a signal to the Electronic Control Module to modify the fuel mixture for idle or full throttle operating conditions.

Idle speed control

The idle speed control motor, mounted on the throttle body, receives signals from the Electronic Control Module to change the position of the idle throttle stop as changes in operating parameters warrant, such as for high engine warm-up idle speed, idle speed compensation for engine load factors such as air conditioning, transmission range selection, etc. In addition the ECM uses the idle speed motor as a dashpot mechanism to prevent sudden closing of the throttle and attendent high emissions output.

The idle air control valve is mounted on the throttle body, where it controls the amount of air passing through the idle circuit

Manifold absolute pressure sensor

The manifold absolute pressure sensor measures the pressure inside the intake manifold and sends this signal to the Electronic Control Unit, where it is compared to the pressure in the fuel system in order to maintain a constant 10.5 psi differential between the fuel system and the intake manifold.

Barometric pressure sensor

The barometric pressure sensor measures the outside (atmospheric) air pressure, and when this signal is compared to the intake manifold pressure, air temperature and engine rpm, the signal can be interpreted by the electronic control unit as the flow rate of air entering the engine.

Electronic control unit

The Electronic Control Unit (ECU) is the computerized control system for the fuel injection, as well as for the ignition timing. Signals are received at the ECM from the coolant temperature sensor, intake manifold air temperature sensor, manifold absolute pressure sensor, outside air pressure sensor, ignition system (rpm) and throttle position sensor. After receiving and processing these signals, the ECM sends controlling orders to the fuel pump, idle speed control motor, ignition system (ignition timing) and fuel injectors.

Fuel pump

An electric twin-turbine fuel pump is located in the fuel tank, integral with the fuel level sender. The fuel pump is controlled by the Electronic Control Unit, which turns the pump on when the ignition switch is turned on. However, if the engine is not cranked by the starter within one second of the ignition coming on, the pump is turned off by the ECU. *Note: For fuel pressure relief and removal and installation procedures, refer to the General Motors Throttle Body Injection section.*

11 Sequential fuel injection (SFI)

General information

The General Motors sequential fuel injection system is used on engines equipped with turbochargers. It utilizes an injector solenoid at each intake port.

The injector is a standard electrically-operated solenoid. Fuel pressure in the system is maintained at 28-to-36 psi and the Electronic Control Module (ECM) controls the injector, the opening time controlling the amount of fuel delivered to the engine. Each injector receives a signal to inject the required amount of fuel just prior to the opening of the intake valve. This gives better control of the fuel mixture than is possible with either throttle body injection, where the fuel is injected far upstream from the intake valve, or multi-port fuel injection, which operates all the injectors at the same time, shooting half the required fuel at the intake valve each time the crankshaft comes around, even though the intake valve opens only every other revolution.

General Motors

12 Sequential fuel injection (SFI) components

Electronic control module

In determining how much fuel is required at any given moment, the Electronic Control Module processes signals from sensors recording engine coolant temperature, exhaust oxygen content, throttle position, intake air mass, engine rpm, vehicle speed and accessory load.

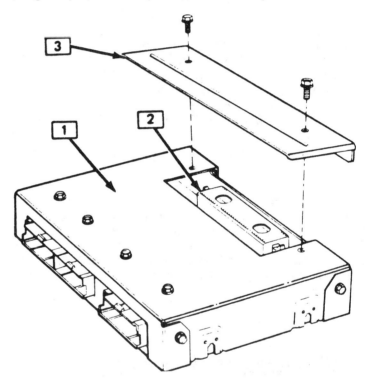

The Electronic Control Module used with the sequential fuel injection system processes the signal from the air mass sensor, as well as other engine sensors, to determine how long the injectors should be held open for optimum performance

1 Electronic control module
2 PROM
3 PROM cover

Air mass sensor

The intake air mass sensor measures the mass (weight) of the air entering the intake manifold, rather than the volume, which gives better control of the air/fuel mixture. The mass of any given volume of air is dependent upon the temperature of the air. Cold air weighs more (has more mass) than warm or hot air, and therefore the colder the air is, the more fuel is required to maintain the proper fuel/air ratio.

On the Bosch air mass sensor system a platinum wire is used to measure the intake air mass. On the General Motors system, however, a heated film is used. The temperature of the incoming air is constantly monitored, and the air mass sensor plate is maintained at exactly 75 degrees above the temperature of the air. Since the incoming air passing over the plate acts to cool the plate, the measurement of the amount of electrical energy needed to maintain that 75 degree temperature differential accurately indicates the mass of air entering the engine, and therefore the amount of fuel necessary to provide the proper fuel/air ratio.

Sequential

Idle air control

The idle speed on the SFI is also controlled by the ECM, through an idle air control bypass channel. When the engine is cold or when accessory loads, such as air conditioning, necessitate a higher idle speed, the ECM, through a stepper motor, opens the bypass pintle valve to provide more air into the system. At the same time the ECM increases the fuel injector open time to maintain the proper fuel/air ratio.

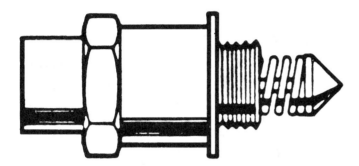

An idle air control valve, set by signal from the Electronic Control Module, controls the idle speed

Fuel pump

While the open time of the injectors is controlled by the ECM, the actual fuel delivery is determined by the fuel pressure, which must be maintained at a constant level. This is accomplished by a positive displacement roller vane pump mounted in the fuel tank. The pump intake is mounted in a special sump in the fuel tank so that a low fuel level, combined with hard cornering, acceleration or deceleration, cannot uncover the pump inlet, causing a pressure drop.

A fuel pressure regulator maintains the required pressure in the system by returning unneeded fuel to the tank. A fuel accumulator is used to dampen the pulses which would otherwise be set up in the fuel rail and fuel line when the injectors open and close.

13 Relieving SFI fuel system pressure

Before disconnecting any components of the fuel system the pressure in the system must be relieved. Note that this pressure can remain in the system long after the engine has been shut off.

To relieve the system pressure, first remove the fuel pump fuse from the fuse block, disabling the fuel pump. Start the engine and let it run until it dies from lack of fuel, then turn the engine over several times with the starter to eliminate all pressure in the system. Be sure to replace the fuel pump fuse after completing work on the system. **Note:** *For removal and installation procedures, refer to the General Motors Port Fuel Injection section.*

Ford Fuel Injection Systems

Central Injection System

Central fuel injection 1
    General description
    Troubleshooting

Central fuel injection components 2
    Electronic Control Assembly (ECA)
    Throttle position sensor
    Oxygen sensor
    Coolant temperature sensor
    Manifold and outside barometric pressure sensors
    Intake charge temperature sensor
    EGR valve position sensor
    Crankshaft position sensor
    Throttle body
    Fuel pressure regulator
    Fuel injectors

Relieving CFI fuel system pressure 3

Typical CFI system removal and installation procedures 4

Port Injection System

Port fuel injection 5
    General description
    Troubleshooting

Port fuel injection components 6
    Air flow meter
    Throttle body assembly
    Air bypass valve
    Oxygen sensor
    Fuel injectors

Relieving port injection fuel system pressure 7

Typical port injection system removal and installation procedures 8

1 Central fuel injection

General description

The Ford Central Fuel Injection system (CFI), is a throttle body type system, with two fuel injectors mounted in a throttle body. The fuel metering is accomplished by modulation of the time pulse sent to the solenoid-type injectors by the Electronic Control Assembly (ECA), which receives signals from a variety of sensors, including an intake charge temperature sensor, manifold and outside barometric pressure sensors, a crankshaft position sensor, coolant temperature sensor, exhaust oxygen content sensor and a throttle position sensor.

Fuel is delivered to the throttle body by an electric fuel pump and a fuel pressure regulator maintains the delivery pressure at a constant level to regulate the amount of fuel that is injected during the fuel injector open pulse from the ECA. Fuel not needed by the injectors is returned to the fuel tank to keep excess pressure from building in the system.

Troubleshooting

Note: All diagnostic causes listed below relate only to the fuel injection system. Quite often the symptom described may have causes other than in the fuel injection system. As an example, "Pinging sound under load" may well be due to a fault in the cooling system or the ignition timing, although such causes are not listed here.

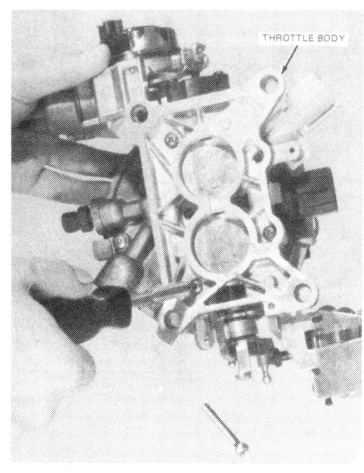

Engine will not start:
Fuel pump not operating
Defective injector

Engine hard to start when cold:
Cold enrichment valve not operating properly
Hot idle compensator stuck open
Intake system vacuum leak

Uneven idle when cold:
Cold enrichment valve not operating properly
Idle setting incorrect
Air cleaner vacuum motor malfunction
Air cleaner duct door malfunction

Stall, stumble or hesitation when cold:
Cold enrichment valve not operating properly
System fuel pressure too low
Fuel filter plugged
Air cleaner vacuum motor malfunction
Air cleaner duct door malfunction

Ford

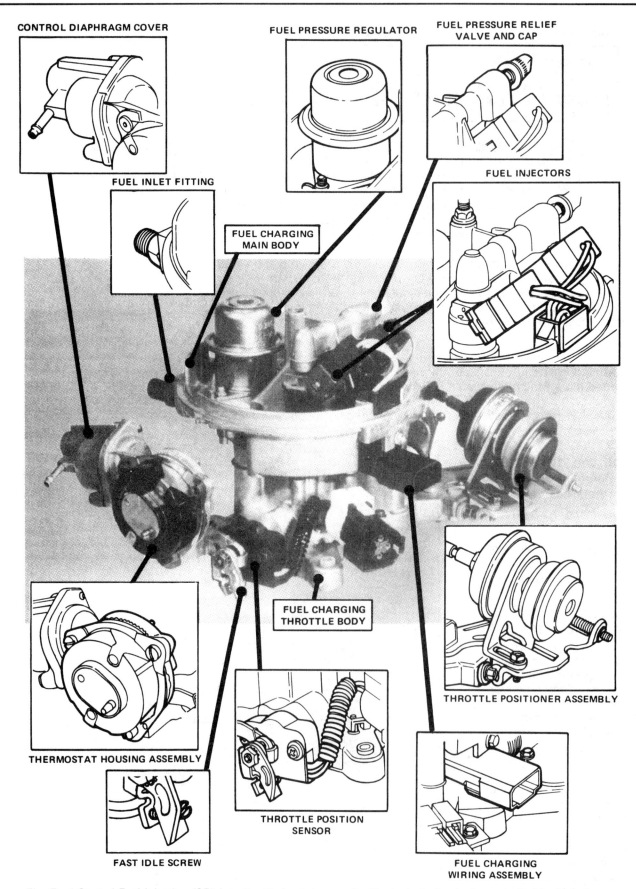

92 *The Ford Central Fuel Injection (CFI) is a throttle body type unit with dual electronically-controlled fuel injectors*

Central (CFI)

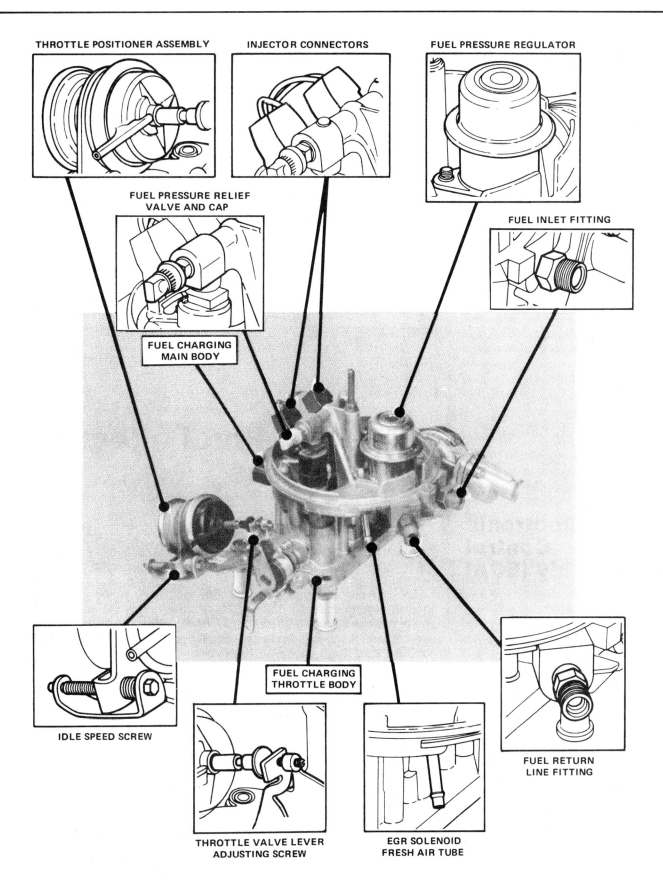

All of the major components of the injection system are mounted on the throttle body, making system repair much easier than with more spread-out systems

Ford

Troubleshooting
(continued)

Engine hard to start when hot:
  Cold enrichment valve not operating properly
  Intake system vacuum leak

Uneven idle when hot:
  Cold enrichment valve not operating properly
  Idle setting incorrect
  Throttle plate sticking
  Hot idle compensator stuck closed

Stall, stumble or hesitation when hot:
  Cold enrichment valve not operating properly
  System fuel pressure too low
  Fuel filter excessively dirty

Stalls during deceleration:
  Idle speed too low
  Throttle position sensor malfunctioning
  Intake system vacuum leak

Loss of power:
  Fuel filter excessively dirty
  Contaminated fuel
  Throttle plate not opening fully
  Throttle linkage sticking
  System fuel pressure too low
  Injectors defective
  Fuel pressure regulator malfunctioning

Surge:
  Fuel filter excessively dirty
  System fuel pressure incorrect
  Fuel pressure regulator malfunction
  Contaminated fuel

2 Central fuel injection components

Electronic Control Assembly (ECA)

The Electronic Control Assembly is a microprocessor which provides reference signals for the engine sensors, receives and interprets the return signals from those sensors, and commands the operation of the fuel injectors (open pulse signal), the ignition timing and the operation of certain emissions system components.

The ECA contains programming information which varies according to the ignition system used (EEC III or EEC IV), the model and engine size, and the emissions system (Federal or California) used. These ECA assemblies are not interchangable, and care should be taken when replacing the unit to make sure the proper ECA is installed.

The Electronic Control Assembly is a microprocessor (computer) which receives signals from various engine sensors and uses those signals to determine the amount of fuel required by the engine at any given moment

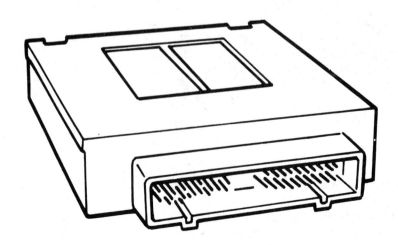

Central (CFI)

Throttle position sensor

The throttle position sensor is mounted on the throttle body assembly. It signals idle, part throttle and full throttle conditions to the electronic control assembly for adjustment of the fuel/air mixture and ignition timing.

The throttle position sensor, mounted on the throttle body, sends a signal to the Electronic Control Assembly to indicate idle, part throttle and full throttle conditions

Oxygen sensor

The oxygen sensor continuously monitors the amount of unburned oxygen in the exhaust stream and signals the electronic control assembly to alter the fuel/air mixture as necessary to maintain the best mixture for both engine performance and emissions control.

Coolant temperature sensor

The coolant temperature sensor initiates a signal to the electronic control assembly when the engine is cold to richen the mixture, retard the ignition timing and modulate the operation of the EGR system during warm-up.

Ford

Manifold and outside barometric pressure sensors

The outside barometric pressure sensor signals variations in pressure, due to temperature or altitude changes, which would alter the amount of air flowing into the engine and therefore the amount of fuel required. The manifold pressure sensor registers variations resulting from changes in engine load, engine speed and throttle opening, sending those signals to the electronic control assembly. The ECU combines the two signals to determine the amount of fuel needed under the current operating conditions.

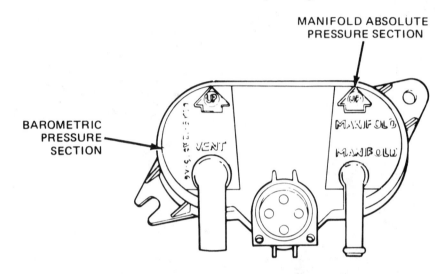

Two pressure sensors are combined in one unit. The barometric pressure sensor measures the outside air pressure, and the MAP sensor monitors the pressure inside the intake manifold

Intake charge temperature sensor

As the temperature of the incoming air changes, the density of that air also changes. In addition, the spray of fuel into the incoming air from the throttle body mounted injectors will also change the temperature of the fuel charge, increasing charge density. Signals recording these changes in density are sent to the electronic control assembly by the intake charge temperature sensor to modulate the amount of fuel injected, maintaining the proper fuel/air mixture for optimum engine performance and minimum emissions.

EGR valve position sensor

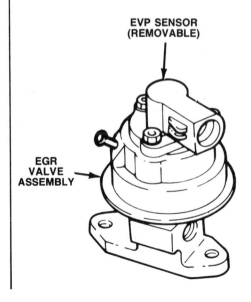

The EGR valve position sensor (EVP) is mounted on the EGR valve. The sensor signals the amount of exhaust gas being routed into the intake system so the Electronic Control Assembly can adjust the fuel/air mixture ratio accordingly

Since a certain amount of exhaust gases are recycled into the intake system for emissions control, a sensor is used to record the position of the EGR valve, and therefore the amount of exhaust allowed into the intake system. The electronic control assembly subtracts this exhaust gas addition from the total flow through the intake system so that only the outside air and fuel are taken into consideration when determining the moment-to-moment fuel requirements of the engine.

Central (CFI)

Crankshaft position sensor

A magnetic pickup on the nose of the crankshaft provides the electronic control assembly with a crankshaft position signal which is used both as an ignition timing reference and to control injector operation.

Throttle body

The throttle body assembly consists of the throttle plate (air valve), a throttle position sensor, an engine idle speed control and an electrically heated bimetallic spring cold idle speed motor. The fuel injectors are mounted in the throttle body, just above the throttle plate.

Fuel pressure regulator

The fuel pressure regulator maintains a constant 39 psi pressure to the injectors, returning excess fuel to the fuel tank to maintain that pressure level.

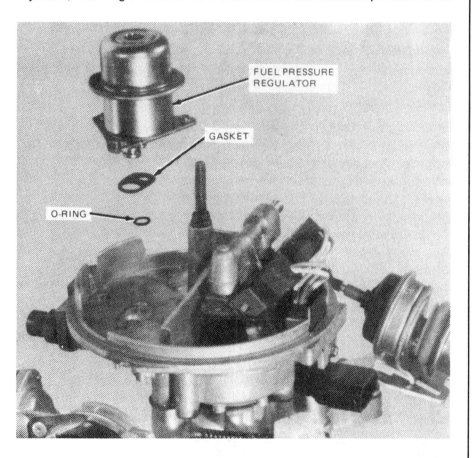

The fuel pressure regulator is mounted on the throttle body, where it maintains a constant pressure in the fuel system, returning excess fuel to the tank

Ford

Fuel injectors

The fuel injectors are electrically operated solenoid valves which are opened by a signal from the electronic control assembly. The fuel pressure at the injector is held at a constant 39 psi by the fuel pressure regulator, so the amount of fuel injected, and therefore the fuel/air ratio, is controlled entirely by the length of time the pintle valve in the injector is held open by the electronic control assembly.

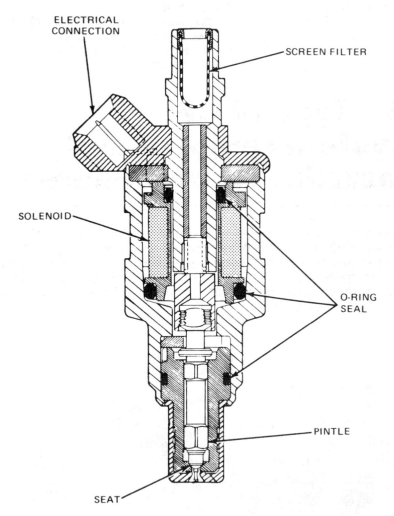

The fuel injectors used in the Ford CFI system are standard Bosch-type electronically-controlled, electrically actuated solenoid units. A signal from the Electronic Control Assembly controls the amount of time the injector is open, and therefore the amount of fuel injected

3 Relieving CFI fuel system pressure

Because pressure is maintained in the system long after the ignition has been shut off, it is necessary to relieve the fuel system pressure before any work can be done on the fuel system.

There are two methods to relieve the pressure in the system. The first requires the use of a hand-operated vacuum pump.

A valve is provided on the throttle body assembly for relieving the fuel pressure. Remove the cap from the fuel tank and disconnect the vacuum hose from the fuel pressure regulator. Connect the hand pump to the fuel pressure regulator and pump it up to approximately 25 in-Hg. At this point the pressure in the system will be released into the fuel tank through the fuel return hose.

An alternative method for fuel pressure relief is to disconnect the electrical connector to the fuel pump and the coil wire from the distributor cap. Be sure to ground the coil wire to prevent arcing.

Turn the engine over for 20-to-30 seconds with the starter, which will eliminate the pressure in the fuel system.

4 Typical CFI system removal and installation procedures

Note: While installation on various models of vehicles produced by Ford varies due to manufacturer installed component location, engine configuration (V-type or inline) and number of cylinders, the basic design of the Ford central fuel injection system will be the same in all cases, and the following typical removal and installation procedure will, with only slight modifications, outline the procedure for your vehicle.

After removing the four screws from the bottom, separate the main body from the throttle body

Removal

1 Remove the air cleaner assembly.
2 Relieve the pressure in the fuel system.
3 Disconnect the throttle cable.
4 If equipped with an automatic transmission, disconnect the transmission throttle valve cable or lever.
5 Disconnect and label all electrical connectors to the throttle body assembly.
6 Disconnect the fuel lines, including the return line.
7 Disconnect and label any vacuum lines connected to the throttle body assembly.
8 Remove the nuts holding the throttle body assembly to the intake manifold and remove the throttle body.
9 Use four 5/16 x 2-1/2 inch bolts with two nuts per bolt to make four legs to stand the throttle body assembly on. Install the bolts in the four intake manifold stud holes in the throttle body flange and use the nuts above and below the flange to secure the bolts.
10 Remove the air cleaner stud.
11 Turn the throttle body assembly upside-down and remove the four Phillips head screws from the bottom of the throttle body.
12 Separate the lower throttle body unit from the upper fuel charging unit.

Removal (continued)

Fuel charging assembly

13. Remove the screws retaining the pressure regulator to the fuel charging assembly and remove the pressure regulator.

14. Disconnect the electrical connectors from the fuel injectors. **Caution:** *Disconnect the electrical connectors by pulling on the connectors, not on the wires.*

15. Loosen but do not remove the wiring harness retaining screw.

16. Push in on the release tabs to remove the wiring harness from the fuel charging assembly.

17. Remove the fuel injector retainer screw and the fuel injector retainer.

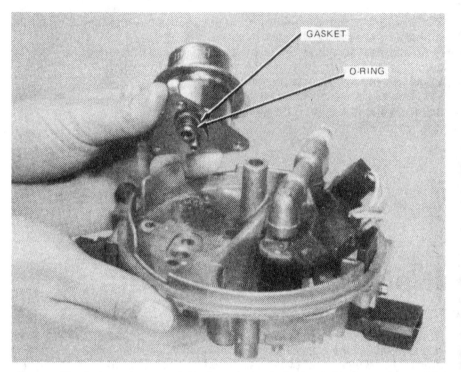

Remove the fuel pressure regulator from the main body, making sure the gasket and O-ring come off with the regulator

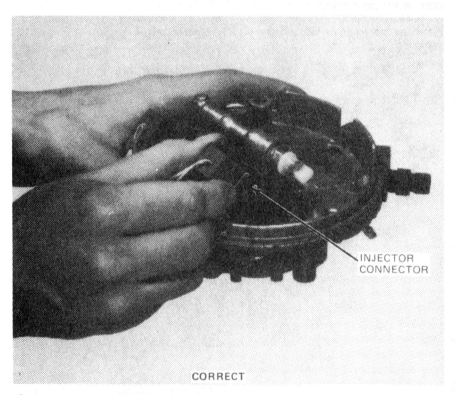

Grasp the connector (not the wires) and pull the connector from the fuel injector

Central (C

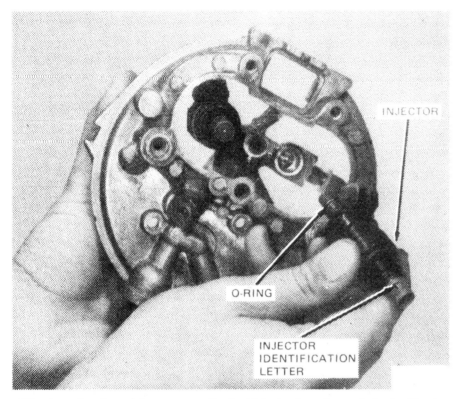

After removing the retainer, remove the fuel injector from the main body. Check to make sure the O-ring remained on the injector, and note the identification letter, as the injector must be installed in the same side from which it was removed

18 Remove the fuel injectors from the fuel charging assembly. Be sure to label the injectors as to which side of the assembly they were removed from. The choke and throttle sides require different injectors. *Note: Each injector is installed in the fuel charging assembly with a small O-ring. Check to make sure that the O-ring comes out with the injector.*

19 Remove the fuel diagnostic valve assembly. This completes the disassembly of the fuel charging assembly.

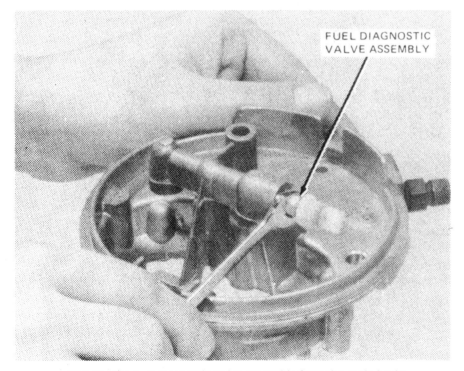

Remove the fuel diagnostic valve assembly from the main body

Throttle body unit

20 Note the position of the index marks on the choke cap housing for proper replacement.

21 Remove the three retaining ring screws and detach the ring, choke cap and gasket.

22 Remove the thermostat lever screw and the thermostat lever.

23 Remove the fast idle cam assembly.

24 Remove the fast idle control rod positioner.

25 Hold the control diaphragm to keep it from moving while removing the two retaining screws.

26 Remove the control diaphragm cover, spring and pulldown control diaphragm.

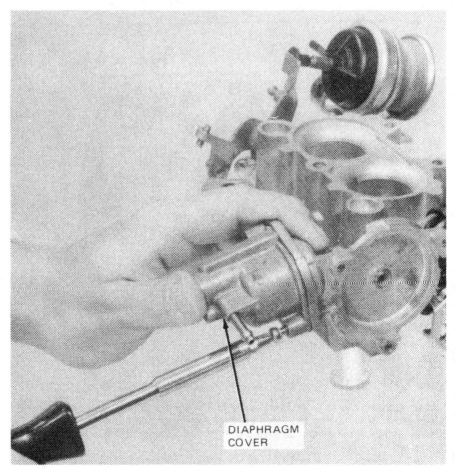

When removing the control diaphragm cover screws from the throttle body hold the cover securely to keep it from moving

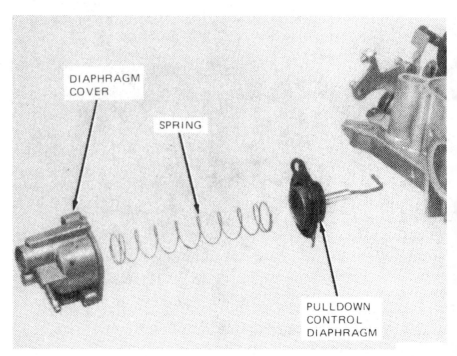

With the screws removed the diaphragm cover, spring and pulldown control diaphragm can be removed

Central (C

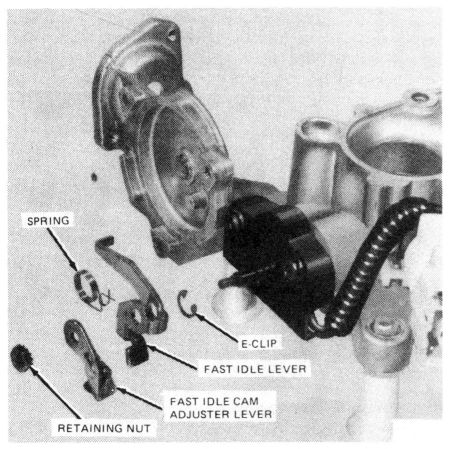

Note the position of the fast idle assembly components as they are removed

After removing the bolts holding the throttle positioner to the throttle body the throttle positioner can be slid off the throttle shaft

27 Remove the fast idle retaining nut.

28 Remove the fast idle cam adjuster lever, fast idle lever, spring and E-clip.

29 Remove the potentiometer connector bracket retaining screw.

30 Remove the throttle position sensor retaining screws and slide the throttle position sensor off the throttle shaft.

31 Remove the throttle positioner retaining screw and the throttle positioner. This completes the disassembly of the throttle body unit.

Installation

Fuel charging assembly

32 Install the fuel pressure diagnostic valve and cap.

33 Lubricate the fuel injector O-rings with oil and install the injectors. Be sure they are installed in the correct sides and push them in lightly with a twisting motion.

34 Put the injector retainer in position and install the screw.

35 Slide the injector wiring harness into the fuel charging assembly, snap the harness tabs into position and tighten the retaining screw.

36 Attach the electrical connectors to the fuel injectors.

37 Lubricate the fuel pressure regulator with light oil, position a new gasket on the regulator and install the regulator in the fuel charging body.

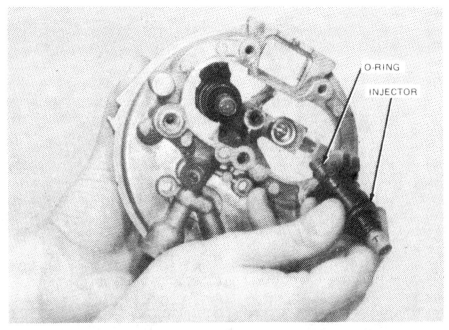

Lubricate the injector O-rings with light oil, check to make sure you have the injectors aligned with the correct sides, then push them into place

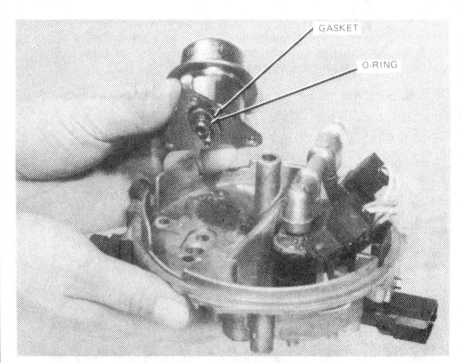

Install a new gasket and O-ring on the fuel pressure regulator, lubricate the O-ring with oil, then install the regulator on the fuel charging body

Central (CFI)

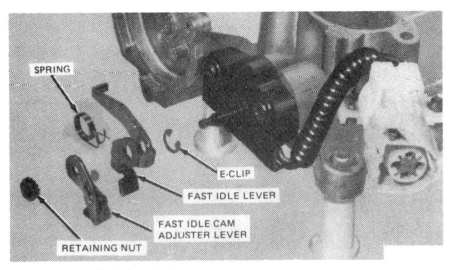

Install the fast idle assembly in the order shown

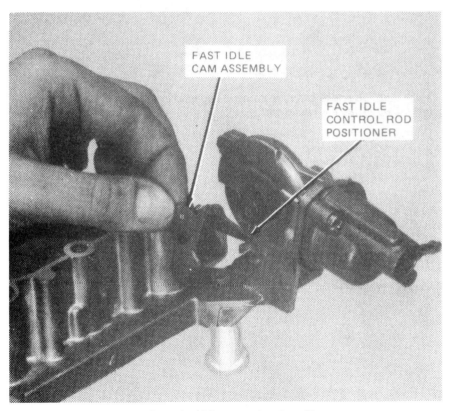

Install the fast idle control rod positioner

Throttle body unit

38 Install the throttle positioner.

39 Hold the throttle position sensor so the wires face up and slide the sensor onto the throttle shaft.

40 Rotate the potentiometer clockwise until it aligns with the screw holes in the throttle body unit and install the screws.

41 Install the potentiometer wiring harness screw.

42 Install the E-clip, fast idle lever and spring, fast idle adjustment lever and fast idle retaining nut.

43 Install the pull down control diaphragm, control modulator spring and diaphragm cover.

44 Hold the diaphragm cover in position and install the two retaining screws.

45 Install the fast idle control rod positioner.

46 Install the fast idle cam.

Ford

Installation
(continued)

47 Install the choke thermostat lever and screw.

48 Install the choke cap with gasket and retaining ring. Be sure the curved end of the bimetallic spring in the choke cap properly engages in the fingers of the thermostat lever.

49 Align the index marks on the choke cap and throttle body and install the retaining screws.

50 Install a new gasket on the fuel charging body and install, but do not tighten, the four screws attaching the fuel charging body to the throttle body unit.

51 Install the air cleaner stud.

52 Tighten the fuel charging body to throttle body screws.

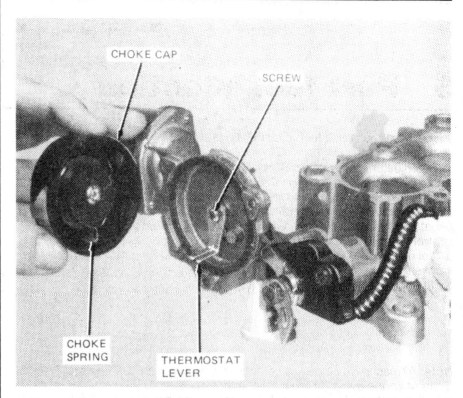

Make sure the choke spring is properly engaged in the slot in the thermostat lever, then install the choke cap

Multi-port

5 Port fuel injection

General description

The Ford multi-port fuel injection system utilizes an injector solenoid at each intake port.

The injector is a standard Bosch-type electrically-operated solenoid. Fuel pressure in the system is maintained at 39 psi differential between the pressure in the fuel line and the pressure on the manifold side of the fuel injectors under normal operating conditions. The pressure is reduced to 30 psi at idle and under other high manifold vacuum conditions. The EEC-IV control unit controls the injectors, the opening time regulating the amount of fuel delivered to the engine. All the injectors fire simultaneously, once each crankshaft revolution. Pressure is maintained by returning excess fuel to the fuel tank.

Troubleshooting

Note: All diagnostic causes listed below relate only to the fuel injection system. Quite often the symptom described may have causes other than in the fuel injection system. As an example, "Pinging sound under load" may well be due to a fault in the cooling system or the ignition timing, although such causes are not listed here.

Engine will not start:
  Fuel pump defective
  Injector malfunction

Engine hard to start when cold:
  Cold enrichment valve not operating properly
  Hot idle compensator malfunction
  Intake system vacuum leak

Uneven idle when cold:
  Cold enrichment valve not operating properly
  Idle setting incorrect
  Air cleaner vacuum motor malfunction
  Air cleaner duct door malfunction

Stall, stumble or hesitation when cold:
  Cold enrichment valve not operating properly
  System fuel pressure too low
  Fuel filter plugged
  Air cleaner vacuum motor malfunction
  Air cleaner duct door malfunction

Ford

Troubleshooting
(continued)

Engine hard to start when hot:
  Cold enrichment valve not operating properly
  Intake system vacuum leak

Uneven idle when hot:
  Cold enrichment valve not operating properly
  Idle setting incorrect
  Throttle plate sticking
  Hot idle compensator stuck closed

Stall, stumble or hesitation when hot:
  Cold enrichment valve not operating properly
  System fuel pressure too low
  Fuel filter excessively dirty

Stalls during decelleration:
  Idle speed too low
  Throttle position sensor malfunctioning
  Intake system vacuum leak

Loss of power:
  Fuel filter excessively dirty
  Contaminated fuel
  Throttle plate not opening fully
  Throttle linkage sticking
  System fuel pressure too low
  Injectors defective
  Fuel pressure regulator malfunctioning

Surge:
  Fuel filter excessively dirty
  System fuel pressure incorrect
  Fuel pressure regulator malfunction
  Contaminated fuel

6 Port fuel injection components

Air flow meter

The Ford air flow meter utilizes an internal vane connected to a potentiometer to signal the amount of air entering the engine to the Electronic Control Assembly

The Air Flow Meter uses a flap connected to a position sensor (potentiometer) to monitor the amount of air entering the engine. Mounted in conjunction with the air flow flap is an air temperature sensor. Since the density of the air varies according to the temperature, the signal from the air temperature sensor is used to modify the signal from the air flow meter to give the electronic control assembly an accurate reading of the fuel necessary to maintain a proper fuel/air mixture.

Multi-port

The throttle body regulates the air flow into the engine by a throttle plate connected to the throttle pedal. A throttle position sensor on the throttle body signals the throttle position to the electronic control assembly, and an idle adjustment screw is provided for fine idle adjustment.

Throttle body assembly

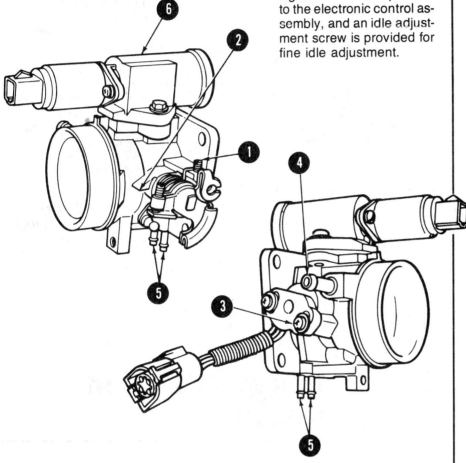

The throttle body assembly used with the multi-port injection system controls the air flow into the engine

1. Minimum idle airflow adjustment screw
2. Wide open throttle stop
3. Throttle position sensor
4. Air source for the PCV system
5. Vacuum taps for the PCV and EGR control signals
6. Idle air bypass valve

The air bybass valve is a solenoid operated valve which meters both cold (outside) and hot (exhaust manifold stove pipe) air into the throttle body in response to commands from the electronic control assembly to maintain an even idle speed. By mixing warm and cold air the mixture density (and therefore the idle speed) can be controlled electronically without movement of the throttle plate.

Air bypass valve

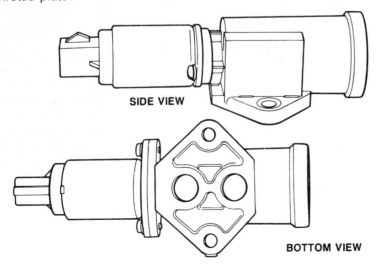

The idle air bypass valve mixes cold outside air with warm air from the exhaust manifold heat stove to control the idle speed

Ford

Oxygen sensor

The oxygen sensor monitors the amount of oxygen in the exhaust stream, determining the required fuel/air ratio at any given moment and signalling the electronic control assembly to change the open time of the injectors as necessary to maintain the correct ratio.

The oxygen sensor allows the Electronic Control Assembly to monitor the fuel/air mixture, constantly adjusting it to provide optimum performance with minimum emissions

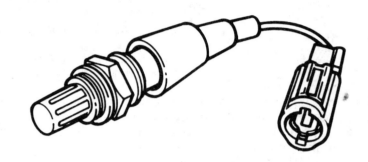

Fuel injectors

The fuel injectors are electrically-operated solenoid valves which are opened by a signal from the electronic control assembly. The fuel pressure at the injector is held at a constant 39 psi by the fuel pressure regulator, so the amount of fuel injected, and therefore the fuel/air ratio, is controlled entirely by the length of time the pintle valve in the injector is held open by the electronic control assembly.

The fuel injectors used in the Ford multi-port system are electronically-controlled, electrically-operated units mounted adjacent to each intake port

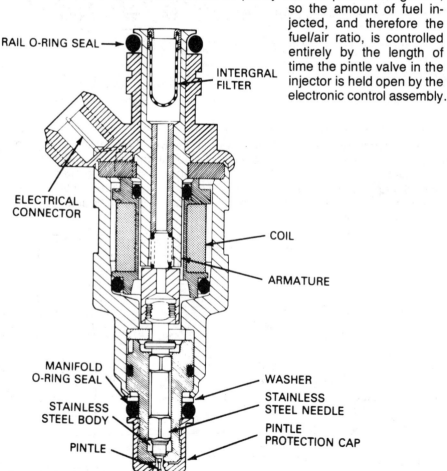

Multi-port

7 Relieving port injection fuel system pressure

Because pressure is maintained in the system long after the ignition has been shut off, the fuel system pressure must be relieved before any work can be done on the fuel system.

There are two methods which may be used to relieve the pressure in the system. The first requires the use of a hand-operated vacuum pump.

A valve is provided on the throttle body assembly for relieving the fuel pressure. Remove the cap from the fuel tank and disconnect the vacuum hose from the fuel pressure regulator. Connect the hand pump to the fuel pressure regulator and pump it up to approximately 25 in-Hg. At this point the pressure in the system will be released into the fuel tank through the fuel return hose.

An alternative method for fuel pressure relief is to disconnect the electrical connector to the fuel pump and the coil wire from the distributor cap. Be sure to ground the coil wire to prevent arcing.

Turn the engine over for 20-to-30 seconds with the starter, which will eliminate the pressure in the system.

8 Typical port injection system removal and installation procedures

Note: *While installation on various models of Ford-produced vehicles varies due to manufacturer installed component location, engine configuration (V-type or inline) and number of cylinders, the basic design of the Ford port fuel injection system will be the same in all cases, and the following typical removal and installation procedure will, with only slight modifications, outline the procedure for your vehicle.*

Removal

1. Drain the coolant from the radiator.
2. Disconnect the negative cable at the battery.
3. Remove the fuel cap to relieve fuel tank pressure.
4. Release the pressure from the fuel system.
5. Disconnect the electrical connector at the throttle position sensor.
6. Disconnect the injector wiring harness.
7. Disconnect the air charge temperature sensor connector.

Label and disconnect the vacuum lines and electrical connectors from the upper intake manifold

Ford

Removal
(continued)

8. Disconnect the engine coolant temperature sensor.
9. Disconnect the vacuum lines at the upper intake manifold vacuum tree. Label all hose connections with tape to aid reinstallation.
10. Disconnect the vacuum line to the EGR valve.
11. Disconnect the vacuum line to the fuel pressure regulator.
12. Remove the throttle linkage shield.
13. Disconnect the throttle linkage.
14. Disconnect the cruise control linkage, if so equipped.
15. If equipped with an automatic transmission, disconnect the kickdown cable.
16. Unbolt the accelerator cable from the bracket and position it out of the way.
17. Disconnect the air intake hose.
18. Remove the air bypass hose.
19. Disconnect the crankcase vent hose.
20. Disconnect the PCV system by detaching the hose from the fitting on the underside of the upper intake manifold.
21. Loosen the hose clamp on the water bypass line at the lower intake manifold and disconnect the hose.
22. Disconnect the EGR tube from the EGR valve by removing the flange nut.
23. Remove the upper intake manifold retaining nuts.
24. Remove the upper intake manifold and air throttle body assembly.
25. Disconnect the push connect fitting at the fuel supply manifold supply line. A special Ford tool is needed to disconnect this type of fitting.
26. Disconnect the return line from the fuel supply manifold.
27. Disconnect the electrical connectors from the fuel injectors and move the harness aside.
28. Remove the fuel supply manifold retaining bolts and carefully remove the fuel supply manifold and injectors.
29. The injectors can be removed from the fuel supply manifold with a slight twisting/pulling motion.

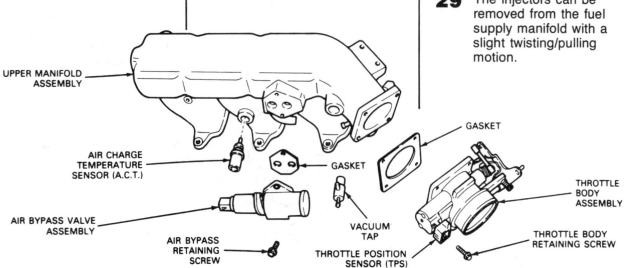

Remove the upper intake manifold, then separate the throttle body and air bypass valve

Multi-port

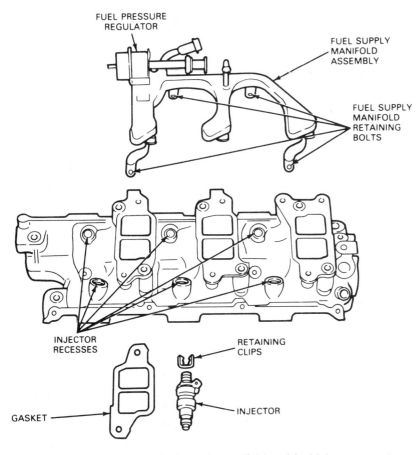

After removing four bolts the fuel supply manifold and fuel injectors can be removed from the lower intake manifold

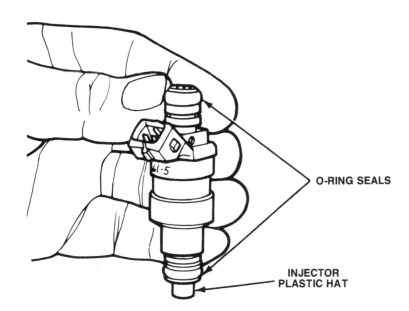

The two O-ring seals should be lubricated with motor oil (not grease) before installation, and check to make sure the plastic injector hat is in place

30 Remove the bottom retaining bolts from the lower manifold.

31 Remove the upper retaining bolts from the lower manifold.

32 Remove the lower intake manifold assembly.

Installation

33 Clean and inspect the mounting faces of the fuel charging manifold assembly and cylinder head.

34 Clean and oil the manifold bolt threads.

35 Install a new gasket.

36 Position the lower manifold assembly on the head and install the engine lift bracket.

37 Install the upper manifold retaining bolts finger tight.

38 Install the remaining manifold bolts.

39 Lubricate new O-rings with motor oil and install two on each injector. *Caution: Do not use silicone grease, as it will clog the injector.*

Ford

Installation
(continued)

40 Using a twisting and pushing motion, install the injectors on the fuel supply manifold.

41 Install the fuel supply manifold and injectors.

42 Connect the electrical connectors to the injectors.

43 Clean the gasket surfaces of the upper and lower intake manifolds.

44 Place a new gasket on the lower intake manifold assembly and place the upper manifold in position.

45 Install the retaining bolts.

46 Connect the supply and return lines to the fuel supply manifold.

47 Connect the EGR tube to the EGR valve and tighten the flange nut.

48 Connect and tighten the water bypass line.

49 Connect the PCV system hose to the fitting on the underside of the upper intake manifold.

50 If the vacuum tree was removed from the upper intake manifold, wrap the threads with Teflon tape before replacing it.

51 Connect the upper intake manifold vacuum hoses.

52 Hold the accelerator cable bracket in position on the upper manifold and install the retaining bolt.

53 Attach the accelerator cable to the bracket.

54 Position a new gasket on the fuel charging assembly air throttle body mounting flange.

55 Attach the air throttle body to the fuel charging assembly and install the retaining nuts and bolts.

56 If equipped with an automatic transmission, connect the kickdown cable.

57 Install the cruise control linkage if so equipped.

58 Install the accelerator cable.

59 Install the throttle linkage shield.

60 Connect the electrical connectors to the throttle position sensor, injector wiring harness, knock sensor, air charge temperature sensor and engine coolant temperature sensor.

61 Reposition the injector wiring harness.

62 Connect the crankcase vent hose.

63 Install the air bypass hose.

64 Install the air intake hose.

65 Connect the negative battery cable.

66 Refill the cooling system.

67 Replace the fuel pressure relief cap, then build up fuel pressure.

68 Without starting the engine, turn the key back-and-forth at least six times between the On and Off positions, leaving the key on for 15 seconds each time. Check the fuel system for leaks.

69 Start the engine and let it run until the coolant temperature stabilizes, then check for leaks.

Chrysler Fuel Injection Systems

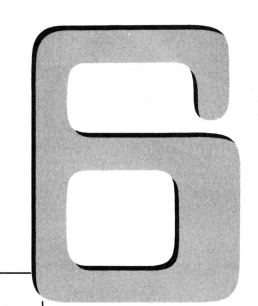

Multi-port System

Multi-port fuel injection 1
    General description
    Troubleshooting

Multi-port fuel injection components 2
    Logic module
    Power module
    Manifold pressure sensor
    Oxygen sensor
    Intake charge temperature sensor
    Coolant temperature sensor
    Detonation sensor
    Throttle body
    Throttle position sensor
    Automatic idle speed control motor
    Fuel pressure regulator
    Fuel injectors

Relieving port injection fuel system pressure 3

Typical port injection system removal and installation procedures 4

Single Point System

Single point fuel injection 5
    General description
    Troubleshooting

Single point fuel injection components 6
    Power module
    Logic module
    Manifold pressure sensor
    Oxygen sensor
    Coolant temperature sensor
    Throttle body temperature sensor
    Throttle body
    Throttle position sensor
    Fuel pressure regulator
    Automatic idle speed control motor
    Fuel injectors

Relieving single point injection system fuel pressure 7

Typical single point system removal and installation procedures 8

1 Multi-port fuel injection

General description

The Chrysler product (Chrysler, Dodge, Plymouth) multi-port fuel injection system is used on turbocharged engines and is controlled by a computer unit called a Logic Module by Chrysler. The logic module is also responsible for emissions control and ignition advance.

The fuel injectors are of the electrical solenoid type, where fuel is delivered to the injectors at a constant 55 psi pressure differential between the fuel side and the manifold side of the injector. The amount of fuel actually injected into the manifold is determined by the length of time the injector is held open by the logic module.

To determine the amount of fuel needed for any given operating condition, the logic module receives signals from a variety of sensors, including a manifold pressure sensor, throttle position sensor, oxygen sensor, coolant temperature sensor, intake charge temperature sensor and vehicle speed sensor.

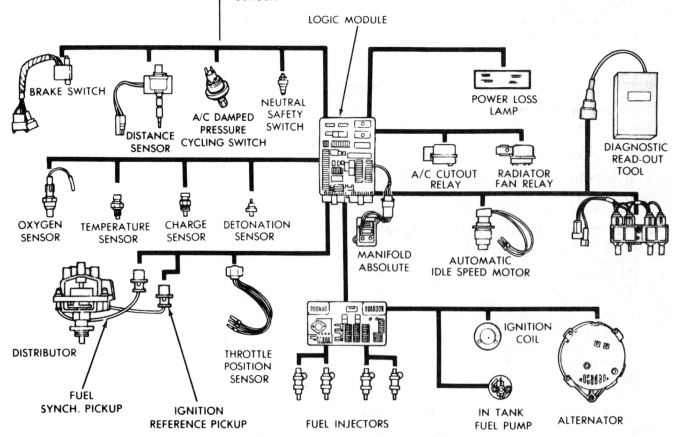

The Chrysler product multi-point fuel injection system utilizes a variety of engine sensors, a logic module to interpret the signals from those sensors, and fuel injectors which control the amount of fuel by the time they are held open by the logic module

Multi-port Troubleshooting

Note: All diagnostic causes listed below relate only to the fuel injection system. Quite often the symptom described may have causes other than in the fuel injection system. As an example, "Engine will not start" may well be due to a fault in the ignition system, although such causes are not listed here.

Engine will not start:
Throttle position switch defective
Faulty logic module
Defective coolant temperature sensor
Low fuel pressure
Intake system air leak
Fuel pump not operating
Faulty power module

Engine hard to start when cold:
Faulty logic module
Malfunctioning coolant temperature sensor
Low fuel pressure
Vacuum leaks
Faulty fuel pump
Faulty throttle position switch

Engine hard to start when hot:
Faulty logic module
Malfunctioning coolant temperature sensor
Low fuel pressure
Excessive evaporative canister purging
Idle speed set too low
Vacuum leaks
Fuel pump malfunctioning

Engine stalls after starting when cold:
Faulty logic module
Faulty coolant temperature sensor
Faulty air temperature sensor
Low fuel pressure
Incorrect idle speed setting
Manifold pressure sensor hose leak

Engine stalls after starting when hot:
Faulty logic module
Low fuel pressure
Improper canister purge system operation
Idle speed set too low
Manifold pressure sensor hose leak or obstruction
Defective, dirty or sticking injectors
Leak in air intake system
Incorrect fuel pressure
Faulty coolant temperature sensor

Rough idle:
Faulty logic module
Manifold pressure sensor hose leak or obstruction
Leak in air intake system

Chrysler

Troubleshooting (continued)

Rough idle (continued)
Incorrect fuel pressure
Faulty coolant temperature sensor
Defective, dirty or sticking injectors

Changes in idle speed:
Throttle position sensor faulty or out of adjustment
Improper canister purge system operation
Throttle body bore contamination
Leak in air intake system
Incorrect fuel pressure
Faulty coolant temperature sensor
Defective, dirty or sticking injectors
Faulty logic module
Contaminated oxygen sensor

Poor driveability when hot:
Faulty logic module
Low fuel pressure
Damaged fuel lines
Manifold pressure sensor hose leak or obstruction
Defective, dirty or sticking injectors
Leak in air intake system
Contaminated oxygen sensor

Hesitation:
Throttle position sensor faulty or out of adjustment
Faulty logic module
Manifold pressure sensor hose leak or obstruction
Incorrect fuel pressure
Leak in air intake system
Improper canister purge system operation

Poor driveability when cold:
Faulty coolant temperature sensor
Faulty air temperature sensor
Low fuel pressure

Misfire on light acceleration:
Faulty logic module
Low fuel pressure
Defective, dirty or sticking injectors

Multi-port Troubleshooting (continued)

Misfire (continued)
Vacuum leak
Unequal fuel delivery between cylinders
Idle speed adjustment incorrect
Throttle position sensor faulty or out of adjustment

Sluggish performance
Faulty logic module
Low fuel pressure
Restricted fuel filter
Damaged fuel lines
Leak in air intake system
Throttle butterfly not opening completely
Throttle position sensor faulty or out of adjustment
Defective, dirty or sticking injectors
Air filter dirty

Intermittent cut out:
Throttle position sensor faulty or out of adjustment
Faulty logic module
Leak in air intake system
Defective, dirty or sticking injectors
Restricted fuel filter
Low fuel pressure

Cuts out under heavy load:
Faulty logic module
Leak in air intake system
Faulty fuel pump
Defective, dirty or sticking injectors

Surge:
Faulty logic module
Incorrect fuel pressure
Restricted fuel filter
Defective, dirty or sticking injectors
Vacuum leak

Excessive fuel consumption
Faulty logic module
Fuel pressure too high
Idle speed adjustment incorrect
Manifold pressure sensor hose leak or obstruction
Vacuum leak
Faulty coolant temperature sensor

Engine runs after ignition is shut off (diesels)
Defective, dirty or sticking injectors
Idle speed adjustment incorrect

2 Multi-port fuel injection components

Logic module

The logic module is the microprocessor control unit (computer) for the fuel injection, ignition and emissions control systems. It receives signals from various sensors and switches, and based on the information from those sources it controls the amount of time the injectors are held open (the fuel/air mixture ratio), as well as the amount the ignition timing is advanced or retarded, and the idle speed.

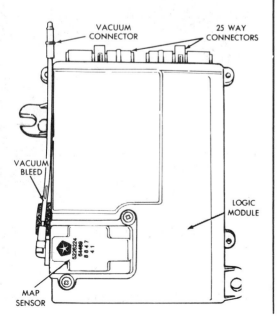

The Chrysler logic module (computer) controls the operation of the fuel injection system, as well as the ignition and emissions control systems

Chrysler

Power module

In order to isolate the engine operating systems (ignition coil, fuel injection system and alternator) from the passenger compartment systems, and to maintain a constant 8-volt (not 12-volt battery voltage) power supply to the logic module, a power module is used between the power source (the battery) and the operating systems. The power module also contains the Automatic Shut Down Relay, which controls the fuel pump, fuel injector power supply, ignition coil and parts of the logic module.

The power module is used to provide a constant source of 8-volt current to the logic module and the various engine sensors

Manifold pressure sensor

The manifold pressure sensor transmits a signal to the logic module regarding manifold vacuum levels (or manifold pressure where the turbocharger is supplying boost to the engine). Along with other signals, the pressure signal is used by the logic module to determine engine load, and therefore the amount of fuel required for the current operating conditions.

The manifold pressure sensor is used by the logic module to determine engine load conditions, and therefore the amount of fuel required

Multi-port

Oxygen sensor

The oxygen sensor is mounted in the exhaust manifold, where it measures the exhaust gas oxygen content to determine whether the fuel/air mixture ratio is too lean or too rich. It sends a signal to the logic module to adjust the mixture as necessary to maintain a proper ratio.

The amount of unburned oxygen in the exhaust stream is monitored by the oxygen sensor, and the signal from the sensor to the logic module determines how much fuel is necessary to maintain a proper fuel/air mixture ratio

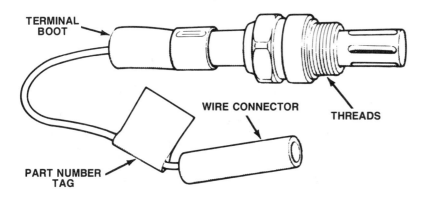

Intake charge temperature sensor

The intake charge temperature sensor sends a signal to the logic module regarding the charge density passing through the intake manifold. The logic module adjusts the amount of fuel injected based on the temperature density of the incoming air supply.

The intake charge sensor measures the temperature of the incoming air and the logic module adjusts the amount of fuel injected to compensate for changes in air density

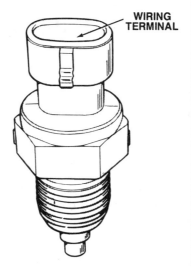

CHARGE SENSOR

Coolant temperature sensor

The coolant temperature sensor signals the logic module when the engine is cold (requiring a richer fuel mixture for proper operation during the warm-up period).

The coolant temperature sensor informs the logic module when the engine is cold, and the logic module richens the mixture and opens the idle air bypass valve to provide a higher idle speed

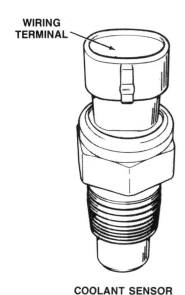

COOLANT SENSOR

Chrysler

Detonation sensor

The detonation sensor monitors the vibrations produced by detonation (spark knock). It sends a signal to the logic module, which alters the ignition timing, boost pressure and fuel/air ratio as necessary to eliminate the detonation condition.

The knock detector (detonation sensor) signals the logic module when detonation is occurring in the engine, and the logic module them alters the ignition timing, fuel/air ratio and the turbo boost to eliminate the detonation condition

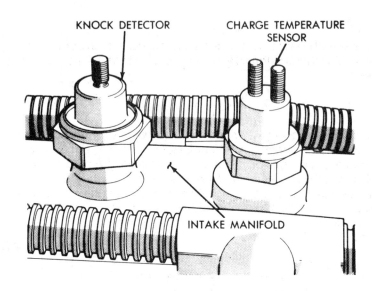

Throttle body

The throttle body controls the airflow from the turbocharger to the intake manifold through a throttle-operated butterfly valve. Also incorporated in the throttle body housing are the throttle position sensor and the automatic idle speed control motor.

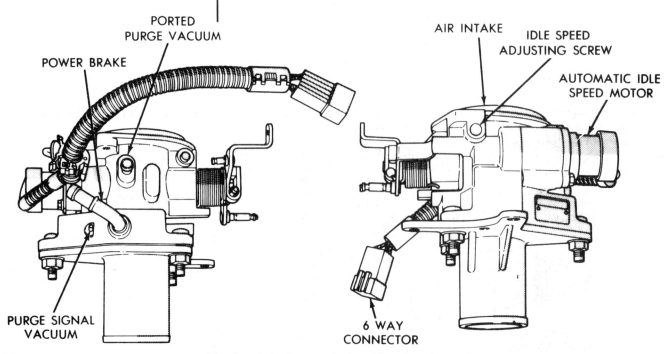

The throttle body controls the airflow into the engine

Multi-port

Throttle position sensor

The throttle position sensor measures the amount the throttle valve is open and transmits a signal to the logic module, where the signal is combined with information from other sensors to adjust the fuel/air ratio for acceleration, idle, wide open throttle operation, etc.

Automatic idle speed control motor

The idle speed control motor is mounted on the throttle body, where it controls the air flow through the idle air bypass to maintain a steady and correct idle speed.

Fuel pressure regulator

The fuel pressure regulator maintains a constant fuel pressure in the system. Because the amount of fuel injected into the engine is controlled by the time the fuel injector is open, the fuel pressure differential between the fuel side of the injector and the manifold must be kept constant. A pressure sensor in the regulator determines the intake manifold pressure and holds the fuel pressure at 55 psi over that level. The pressure is maintained by returning excess fuel to the tank through a fuel return port and return line.

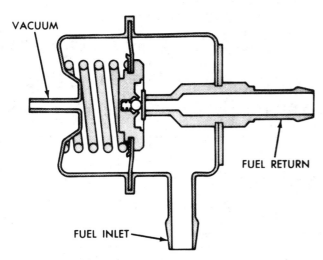

The fuel pressure regulator maintains a constant pressure differential between the fuel system and the intake manifold so that only the length of time the injector is held open determines the amount of fuel injected

Fuel injectors

The fuel injectors are electrical solenoids, with a pintle valve opened by the solenoid when the signal is received from the logic module. The pintle is either open or closed, with no modulated positions. Therefore, the amount of fuel passed through the injector is determined by the time (pulse width) the injector is held open by the logic module.

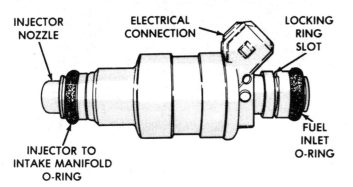

The fuel injector receives an electrical signal from the logic module, at which time a solenoid inside the injector opens the fuel valve, allowing fuel to spray into the intake manifold

Chrysler

3 Relieving port injection fuel system pressure

The fuel system is under pressure even after the engine has been shut off, so the pressure in the system must be relieved before any work is done which requires disconnecting fuel lines or removing fuel injection components.

Remove the fuel tank cap to release any pressure in the tank. Remove the wiring connector from any injector and ground one terminal of the injector with a jumper wire. Use another jumper wire to connect the other injector terminal to the positive battery post for no more than ten seconds. This will release any pressure in the system.

4 Typical port injection system removal and installation procedures

Note: While installation on various models of Chrysler-built vehicles varies due to manufacturer installed component location, engine configuration (V-type or inline) and number of cylinders, the basic design of the Chrysler multi-port fuel injection system will be the same in all cases, and the following typical removal and installation procedure will, with only slight modifications, outline the procedure for your vehicle.

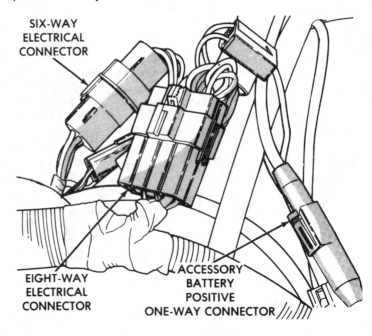

Removal

1. Disconnect the negative cable from the battery.
2. Remove the air cleaner-to-throttle body screws, loosen the hose clamp and remove the air cleaner adapter.
3. Disconnect the 6-way electrical connector at the throttle body.
4. Remove the screws holding the throttle position sensor to the throttle body.
5. Unclip the wiring clip from the convoluted tube and remove the throttle position sensor mounting bracket.
6. Lift the throttle position sensor off the throttle shaft and remove the O-ring.
7. Pull the three wires of the throttle position sensor from the convoluted tube.
8. Lift the locking tabs inside the 6-way connector and remove the blade terminals, noting their positions for future reassembly.
9. Remove the screws holding the automatic idle speed control motor assembly to the throttle body. *Caution: Do not remove the clamp on the automatic idle speed motor or damage to the motor will result.*

Multi-port

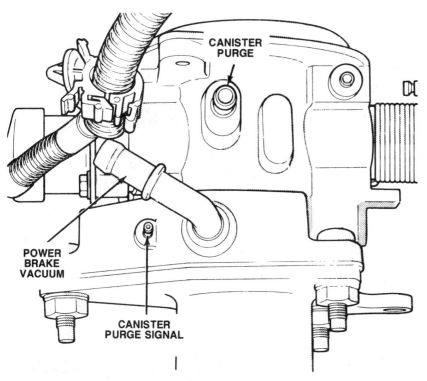

Label and disconnect the various vacuum hoses attached to the throttle body

Remove the throttle cable cruise control linkage (if equipped), and kickdown linkage

10 Lift the locking tabs inside the 6-way connector and remove the two blade terminals from the idle speed motor, noting their positions for future reassembly.

11 Remove the automatic idle speed control motor from the throttle body, making certain that both O-rings remain on the idle speed motor.

12 Remove the throttle cable.

13 If equipped, remove the cruise control linkage.

14 If the vehicle has an automatic transmission, remove the kickdown cable from the throttle linkage.

15 Disconnect and label any vacuum hoses attached to the throttle body assembly.

16 Loosen the throttle body-to-turbocharger hose clamp.

17 Remove the nuts holding the throttle body to the intake manifold and detach the throttle body.

18 Loosen the fuel inlet hose clamp at the fuel rail and remove the fuel hose.

19 Disconnect the fuel pressure regulator vacuum line where it connects to the intake manifold vacuum fitting.

20 Remove the two fuel pressure regulator bracket bolts.

21 Loosen the hose clamp on the hose running from the fuel rail to the fuel pressure regulator and remove the hose and regulator.

22 Disconnect the fuel injector wiring harness at the connector.

23 Remove the fuel injector heat shield clips.

24 Remove the bolts holding the fuel rail to the intake manifold.

Chrysler

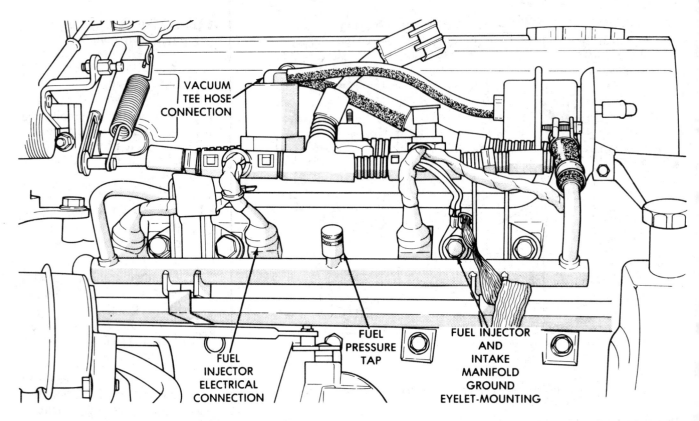

Label and disconnect the vacuum lines and electrical connectors at the fuel rail and fuel pressure regulator

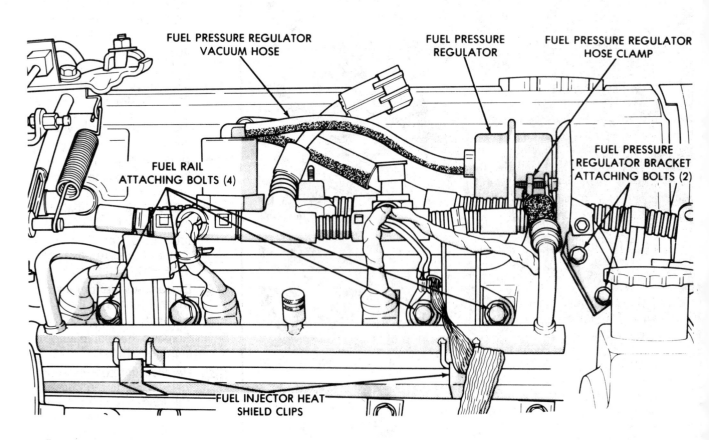

Remove the fuel line to the fuel pressure regulator and the bolts attaching the fuel rail to the intake manifold

Multi-port

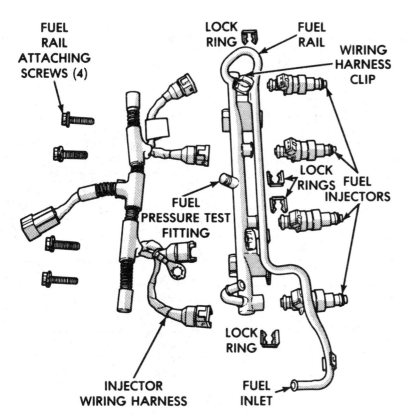

After unplugging the injector wiring harness from the injectors the injector locking clips can be released and the injectors removed from the fuel rail

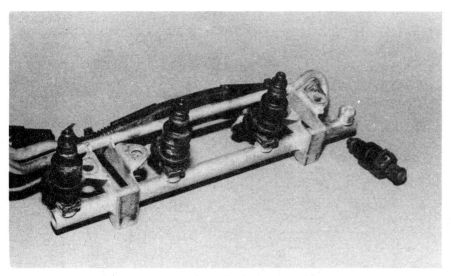

Position the fuel injectors on the fuel rail then install the locking rings

Removal
(continued)

25 Remove the fuel rail and injectors by pulling the injectors straight out of the intake manifold ports.

26 Disconnect the electrical connectors from the fuel injectors.

27 Remove the injector locking clips from the fuel rail and pull the injectors out of the rail.

Installation

28 Lubricate the injector O-ring with light oil and install the injector into the fuel rail.

29 Install the locking clip by sliding the open end into the top slot of the injector and onto the receiver cup ridge.

30 Repeat the procedure for the remaining injectors.

31 Install the injectors into the intake manifold and tighten the four attaching bolts a little at a time to draw the injectors tightly and evenly into the manifold. Be sure to replace any ground straps which may have been mounted between the fuel rail bolts and the intake manifold.

32 Install the electrical connectors onto the injectors.

33 Install the injector heat shield clips.

34 Connect the injector wiring harness to the main harness.

35 Connect the fuel pressure regulator vacuum hose to the manifold vacuum fitting.

Chrysler

36. Install the fuel hose from the fuel pressure regulator onto the fuel rail and tighten the clamp.
37. Install the fuel pressure regulator bracket bolts.
38. Attach the fuel inlet hose to the fuel rail.
39. Install the throttle body on the intake manifold.
40. Install the turbocharger-to-throttle body hose and tighten the clamp.
41. Connect any removed vacuum hoses to the throttle body.
42. Install the throttle cable bracket.
43. If equipped with an automatic transmission, install the kickdown cable.
44. Install the cruise control linkage, if so equipped.
45. Install the throttle cable.
46. Install the automatic idle speed control motor onto the throttle body.
47. Route the automatic idle speed control motor wiring to the 6-way connector and reinstall the blade connectors into their correct positions.
48. Install the throttle position sensor blade terminals in the 6-way connector.
49. Insert the wires from the throttle position sensor into the convoluted tube.
50. Install the throttle position sensor and O-ring, along with the mounting bracket, onto the throttle body.
51. Install the wiring clips on the convoluted tube.
52. Connect the 6-way connector.
53. Install the air cleaner adapter and tighten the hose clamp.
54. Reconnect the negative battery cable.

Install the throttle cable and (if equipped) the cruise control cable

5 Single point fuel injection

Single point

General description

The Chrysler product (Chrysler, Dodge, Plymouth) single point fuel injection is a throttle body type, with a single injector mounted in the throttle body housing. It controlled by a computer unit called a Logic Module by Chrysler. The logic module is also responsible for emissions control and ignition advance.

The fuel injector is an electrical solenoid type, where fuel is delivered to the injector at a constant 14.5 psi pressure differential between the fuel side and the manifold side of the injector. The amount of fuel actually injected into the manifold is determined by the length of time the injector is held open by the logic module.

To determine the amount of fuel needed for any given operating condition, the logic module receives signals from a variety of sensors, including a manifold pressure sensor, throttle position sensor, oxygen sensor, coolant temperature sensor, intake charge temperature sensor and vehicle speed sensor.

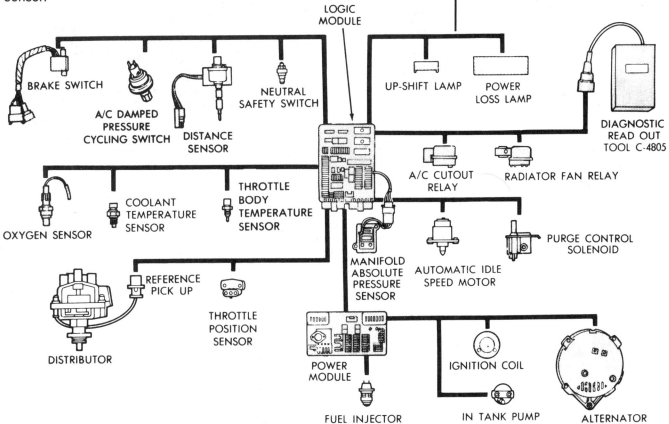

The Chrysler single-point (throttle body) injection system uses a single centrally-mounted fuel injector controlled by signals from the logic module

Chrysler

Troubleshooting

Note: All diagnostic causes listed below relate only to the fuel injection system. Quite often the symptom described may have causes other than in the fuel injection system. As an example, "Engine will not start" may well be due to a fault in the ignition system, although such causes are not listed here.

Engine will not start:
Throttle position switch not working properly
Logic module defective
Coolant temperature sensor malfunction
System fuel pressure too low
Fuel pump not operating
Faulty power module

Engine stalls after starting when cold:
Logic module defective
Coolant temperature sensor faulty
System fuel pressure too low
Incorrect idle speed setting
Manifold pressure sensor malfunction

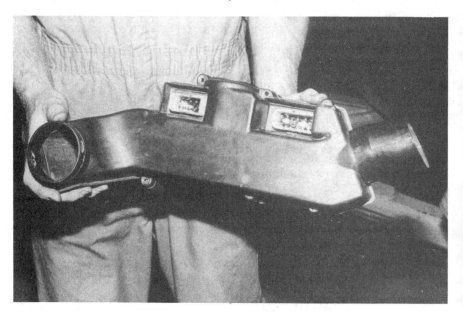

Engine hard to start when cold:
Logic module defective
Coolant temperature sensor defective
System fuel pressure incorrect
Intake system vacuum leak
Faulty fuel pump
Throttle position switch sticking

Engine hard to start when hot:
Logic module defective
Coolant temperature sensor faulty
System fuel pressure too low
Excessive evaporative canister purging
Idle speed set too low
Intake system vacuum leak
Fuel pump malfunctioning

Engine stalls after starting when hot:
Logic module defective
System fuel pressure too low
Improper canister purge system operation
Idle speed set too low
Manifold pressure sensor malfunction
Injector malfunction
System fuel pressure incorrect
Coolant temperature sensor faulty

Rough idle:
Logic module defective
Manifold pressure sensor malfunction
System fuel pressure incorrect
Coolant temperature sensor faulty
Injector malfunction

Single point

Changes in idle speed:
Throttle position sensor faulty or out of adjustment
Improper canister purge system operation
Throttle plate sticking
System fuel pressure incorrect
Coolant temperature sensor faulty
Injector malfunction
Logic module defective
Contaminated oxygen sensor

Poor driveability when cold:
Coolant temperature sensor faulty
Faulty air temperature sensor
System fuel pressure too low

Poor driveability when hot:
Logic module defective
System fuel pressure too low
Damaged fuel lines
Manifold pressure sensor malfunction
Injector malfunction
Contaminated oxygen sensor

Hesitation
Throttle position sensor faulty or out of adjustment
Logic module defective
Manifold pressure sensor malfunction
System fuel pressure incorrect
Improper canister purge system operation

Misfire on light acceleration:
Logic module defective
System fuel pressure too low
Injector malfunction
Intake system vacuum leak
Idle speed adjustment incorrect
Throttle positions sensor faulty or out of adjustment

Sluggish performance
Logic module defective
System fuel pressure too low
Restricted fuel filter
Damaged fuel lines
Throttle plate not opening completely
Throttle position sensor faulty or out of adjustment
Injector malfunction
Air filter dirty

Intermittent cut out:
Throttle position sensor faulty or out of adjustment
Logic module defective
Injector malfunction
Restricted fuel filter
System fuel pressure too low

Cuts out under heavy load:
Logic module defective
Faulty fuel pump
Injector malfunction

Surge:
Logic module defective
System fuel pressure incorrect
Restricted fuel filter
Injector malfunction
Intake system vacuum leak

Excessive fuel consumption
Logic module defective
System fuel pressure too high
Idle speed adjustment incorrect
Manifold pressure sensor malfunction
Intake system vacuum leak
Faulty coolant temperature sensor

Engine runs after ignition is shut off (diesels)
Injector malfunction
Idle speed adjustment incorrect

Chrysler

6 Single point fuel injection components

Power module

In order to isolate the engine operating sytems (ignition coil, fuel injection system and alternator) from the passenger compartment systems, and to maintain a constant 8-volt (not 12-volt battery voltage) power supply to the logic module, a power module is used between the power source (the battery) and the operating systems. The power module also contains the Automatic Shut Down Relay, which controls the fuel pump, fuel injector power supply, ignition coil and parts of the logic module.

To separate the vehicle electrical system from the fuel and ignition control systems, a separate 8-volt power supply is provided

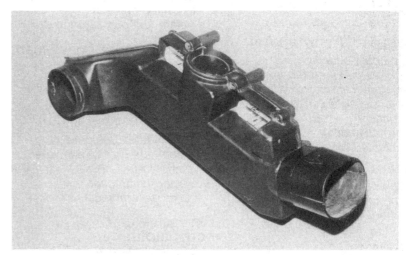

Logic module

The logic module is the microprocessor control unit (computer) for the fuel injection, ignition and emissions control systems. It receives signals from various sensors and switches. Based on the information from those sources it controls the amount of time the injectors are held open (the fuel/air mixture ratio), as well as the amount the ignition timing is advanced or retarded and the idle speed.

The logic module (computer) receives signals from a variety of engine sensors and processes the data to determine the amount of fuel needed under various operating conditions

Single point

Manifold pressure sensor

The manifold pressure sensor transmits a signal to the logic module regarding manifold vacuum levels. Along with other signals, the pressure signal is used by the logic module to determine engine load, and therefore the amount of fuel required for the current operating conditions.

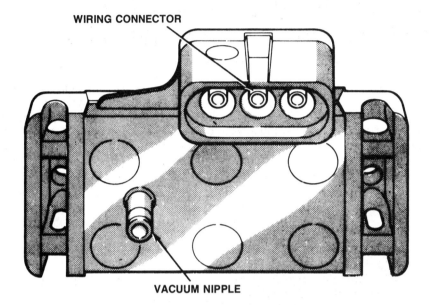

The manifold absolute pressure sensor signals engine-loading conditions to the logic module, which adjusts the amount of fuel being injected accordingly

Oxygen sensor

The oxygen sensor is mounted in the exhaust manifold, where it measures the exhaust gas oxygen content to determine whether the the fuel/air mixture ratio is too lean or too rich. It sends a signal to the logic module to adjust the mixture as necessary to maintain a proper ratio.

The amount of fuel which is mixed with the air entering the engine must remain at a set ratio under all operating conditions, and the oxygen sensor, mounted in the exhaust stream, monitors this ratio and signals necessary adjustments to the logic module

Chrysler

Coolant temperature sensor

When the engine is cold a richer mixture is needed, and the coolant temperature sensor provides the input to the logic module to control the idle speed

The coolant temperature sensor signals the logic module when the engine is cold (requiring a richer fuel mixture for proper operation during the warm-up period).

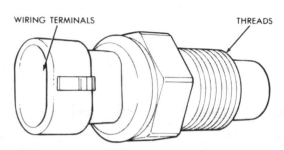

Throttle body temperature sensor

The throttle body temperature sensor monitors the fuel temperature in the throttle body and sends a signal to the logic module to alter the fuel/air mixture ratio to allow for hot restart conditions.

Throttle body

The throttle body contains the throttle plate itself, as well as the fuel injector, the fuel pressure regulator, throttle position sensor, automatic idle speed control motor and throttle body temperature sensor.

Throttle position sensor

The throttle position sensor measures the amount the throttle valve is open and transmits a signal to the logic module, where the signal is combined with information from other sensors to adjust the fuel/air ratio for acceleration, idle, wide open throttle operation, etc.

Fuel pressure regulator

The fuel pressure regulator maintains a constant fuel pressure in the system. Because the amount of fuel injected into the engine is controlled by the time the fuel injector is open, the fuel pressure differential between the fuel side of the injector and the manifold must be kept constant. A pressure sensor in the regulator determines the intake manifold pressure, and holds the fuel pressure at 14.5 psi over that level. The pressure is maintained by returning excess fuel to the tank through a fuel return port and return line.

The fuel pressure regulator maintains the pressure in the fuel system at a set level above the pressure in the intake manifold

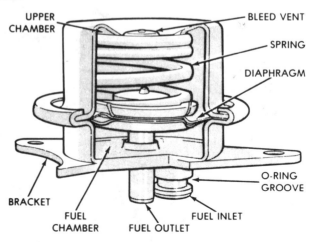

Single point

Automatic idle speed control motor

The idle speed control motor is mounted on the throttle body, where it controls the air flow through the idle air bypass to maintain a steady and correct idle speed.

The automatic idle speed motor is mounted on the throttle body, where it controls the supplementary air flow through the throttle body when a fast idle is needed

Fuel injector

The fuel injector is an electrical solenoid, with a spring loaded ball valve opened by the solenoid when the signal is received from the logic module. The ball valve is either open or closed, with no modulated positions. Therefore, the amount of fuel passed through the injector is determined by the time (pulse width) the injector is held open by the logic module.

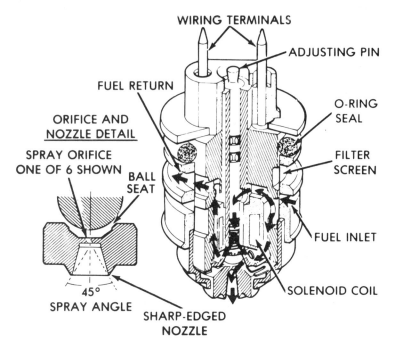

The fuel injector delivers a fine spray of fuel into the intake manifold whenever the valve is opened by the logic module. Since the fuel pressure is maintained at a constant level above intake manifold pressure, the amount of fuel injected is controlled precisely by the length of time the injector is held open

135

Chrysler

Removal

1. Remove the air cleaner.
2. Release the pressure in the fuel system.
3. Disconnect the negative battery cable at the battery.
4. Disconnect and label the vacuum hoses attached to the throttle body.
5. Disconnect the throttle body electrical connectors.
6. Remove the throttle cable.
7. Remove the cruise control linkage if so equipped.
8. If equipped with an automatic transmission, remove the kickdown cable.
9. Remove the throttle return spring.
10. Remove the fuel inlet and return hoses.
11. Remove the throttle body mounting bolts and detach the throttle body from the intake manifold.
12. Remove the three screws attaching the fuel pressure regulator to the throttle body and pull the pressure regulator out of the body.
13. Remove the O-ring from the pressure regulator and detach the gasket.
14. Remove the Torx screw holding down the injector cap.

7 Relieving single point injection system fuel pressure

The fuel system is under pressure even after the engine has been shut off, so the pressure in the system must be relieved before any work is done which requires disconnecting fuel lines or removing fuel injection components.

Remove the fuel tank cap to release any pressure in the tank. Remove the wiring connector from the injector and ground one terminal of the injector with a jumper wire. Use another jumper wire to connect the other injector terminal to the positive battery post for no more than ten seconds. This will release any pressure in the system.

8 Typical single point system removal and installation procedures

Note: While installation on various models of Chrysler-built vehicles varies due to manufacturer installed component location, engine configuration (V-type or inline) and number of cylinders, the basic design of the Chrysler single point fuel injection system will be the same in all cases, and the following typical removal and installation procedure will, with only slight modifications, outline the procedure for your vehicle.

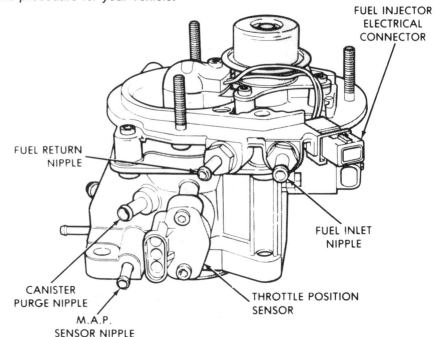

Single point

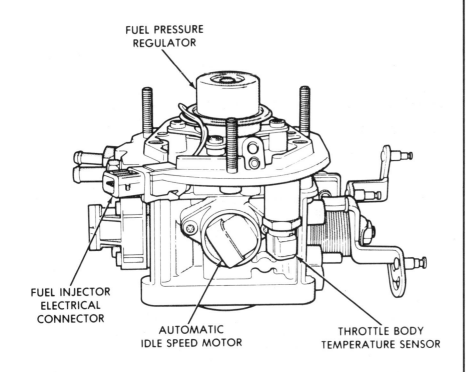

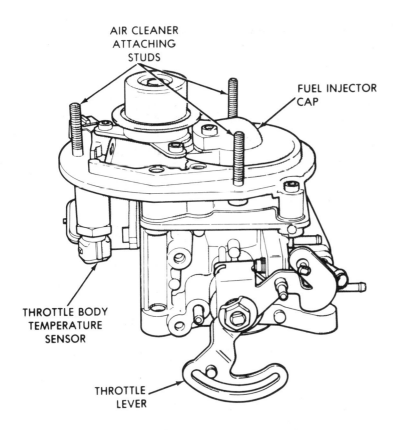

15 With two screwdrivers, lift the cap off the injector using the slots provided.

16 Using a screwdriver placed in the holes in the side of the electrical connector, pry the injector out.

17 Make sure the injector lower O-ring has come out of the throttle body housing along with the injector.

18 Remove the two screws holding the throttle position sensor to the throttle body.

19 Lift the throttle position sensor off the throttle shaft and remove the O-ring.

20 Remove the throttle body temperature sensor by unscrewing it from the throttle body.

21 Remove the two Torx screws attaching the automatic idle speed motor assembly from the throttle body and remove the motor, making sure the O-ring remains with the motor.

Installation

22 Be sure the automatic idle speed motor assembly pintle is in the retracted position.

23 Install the automatic idle speed motor in the throttle body housing, making sure the O-ring is in place.

24 Apply heat transfer grease to the throttle body temperature sensor and install the sensor in the throttle body.

25 Install the throttle position sensor on the throttle shaft, making sure the O-ring is in place.

Chrysler

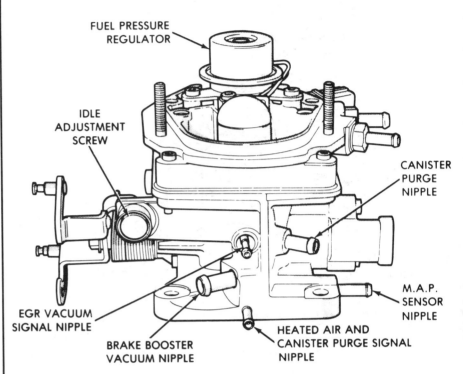

Installation (continued)

26. Place the lower O-ring on the injector and the small O-ring on the injector cap. The upper O-ring should still be in place on the injector.
27. Place the injector in position in the throttle body. Position the injector so that the cap can be installed without interference. The cap and the injector are keyed for proper positioning.
28. Rotate the cap and injector to line up the attachment hole.
29. Push down on the cap to ensure a good seal.
30. Install the injector cap.
31. Place a new gasket on the fuel pressure regulator and make sure the O-ring is in place.
32. Position the pressure regulator on the throttle body and press it into place. Install the screws.
33. Using a new gasket, install the throttle body on the intake manifold.
34. Install the fuel intake and return hoses.
35. Install the throttle return spring.
36. If equipped with an automatic transmission, install the kickdown cable.
37. Install the cruise control linkage, if so equipped.
38. Install the throttle cable.
39. Install the disconnected electrical connectors.
40. Install the disconnected vacuum lines.
41. Install the air cleaner.
42. Reconnect the negative battery cable.

Index

A

Air bypass valve — 109
Air control valve — 17
Air flow sensor
    Ford multi-port — 108
    General information — 12
    K/KE-Jetronic — 53
    L/LH-Jetronic — 36
Air mass sensor
    General information — 12
    GM multi-port — 77
    GM sequential — 88
    Motronic — 36
Applications — 23
Automatic idle speed control motor — 123, 135
Auxiliary air regulator
    General information — 21
    K/KE-Jetronic — 57
    L/LH-Jetronic — 42
    Motronic — 42

B

Barometric pressure sensor
    Ford CFI — 96
    General information — 15
    GM Digital — 87
Boost sensor
    General information — 15
    L/LH-Jetronic — 40
    Motronic — 40
Bosch fuel injection systems — 29

C

Central Fuel Injection (Ford) system
    Barometric pressure sensor — 96
    Coolant temperature sensor — 95
    Crankshaft position sensor — 97
    EGR valve position sensor — 96
    Electronic control assembly — 94
    Fuel injectors — 98
    Fuel pressure regulator — 97
    General description — 91
    Intake charge temperature sensor — 96
    Manifold pressure sensor — 96
    Oxygen sensor — 95
    Relieving fuel pressure — 98
    Removal and installation — 99
    Throttle body — 97
    Throttle position sensor — 95
    Troubleshooting — 91
Chrysler fuel injection systems — 115
Cold start valve
    General information — 20
    GM multi-port — 81
    K/KE-Jetronic — 56
    L/LH-Jetronic 41
    Motronic — 41
Continuous flow injection — 10
Coolant temperature sensor
    Chrysler multi-port — 121
    Chrysler single point — 134
    Ford CFI — 95
    General information — 14
    L/LH-Jetronic — 39
    Motronic — 39

Index

Crankshaft position sensor
  Ford CFI — 97
  General information — 16

D

Detonation sensor
  Chrysler multi-port — 122
  General information — 14
  L/LH-Jetronic — 40
  Motronic — 40
Digital (GM) fuel injection system
  Barometric pressure sensor — 87
  Electronic control unit — 87
  Fuel injectors — 86
  Fuel pressure regulator — 86
  Fuel pump — 87
  General information — 85
  Idle air control — 86
  Manifold pressure sensor — 87
  Throttle body — 85
  Throttle position sensor — 86

E

EGR valve position sensor — 96
Electronic control module (unit)
  Chrysler multi-port — 119
  Chrysler single point — 132
  Ford CFI — 94
  General information — 11, 17
  GM digital — 87
  GM multi-port — 77
  GM sequential — 88
  GM-TBI — 65
  K/KE-Jetronic — 52
  L/LH-Jetronic — 35
  Motronic — 35
Electronically timed injection — 9

F

Ford fuel injection systems — 90
Fuel accumulator
  General information — 22
  K/KE-Jetronic — 55
Fuel distributor
  General information — 22
  K/KE-Jetronic — 54
Fuel injection types — 8
Fuel injectors
  Chrysler multi-port — 123
  Chrysler single point — 135
  Ford CFI — 98
  Ford multi-port — 110
  General information — 18
  GM digital — 86
  GM multi-port — 81, 82
  GM-TBI — 70
  K/KE-Jetronic — 55
  L/LH-Jetronic — 40
  Motronic — 40
Fuel pressure regulator
  Chrysler multi-port — 123
  Chrysler single point — 134
  Ford CFI — 97
  General information — 16
  GM digital — 86
  GM-TBI — 66, 69
  L/LH-Jetronic — 34
Fuel pump
  GM digital — 87
  GM multi-port — 77
  GM sequential — 89
  GM-TBI — 66
  K/KE-Jetronic — 57
  L/LH-Jetronic — 43
  Motronic — 43
Fuel temperature sensor
  General information — 14
  L/LH-Jetronic — 39
  Motronic — 39

G

General Motors fuel injection systems — 62

I

Idle air control valve
  Chrysler multi-port — 123
  Chrysler single point — 135
  GM digital — 86
  GM multi-port — 77, 83
  GM sequential — 89
  GM-TBI — 65, 69, 73
Idle speed bypass valve — 22
Intake charge temperature sensor
  Chrysler multi-port — 121
  Ford CFI — 96
  General information — 15
Introduction — 7

K

K/KE-Jetronic (Bosch) fuel injection systems
  Air flow sensor — 53
  Auxiliary air regulator — 57
  Cold start valve — 56
  Electronic control unit — 52
  Fuel accumulator — 55
  Fuel distributor — 54
  Fuel injectors — 55
  Fuel pump — 57

Index

General information — 50
Line pressure regulator — 54
Oxygen sensor — 55
Relieving fuel pressure — 59
Removal and installation — 59
Thermo-time switch — 56
Throttle valve — 54
Troubleshooting — 51
Warm up regulator — 56

L

L/LH-Jetronic (Bosch) fuel injection systems
Air flow sensor — 36
Auxiliary air regulator — 42
Boost sensor — 40
Cold start valve — 41
Coolant temperature sensor — 39
Detonation sensor — 40
Electronic control unit — 35
Fuel injectors — 40
Fuel pressure regulator — 34
Fuel pump — 43
Fuel temperature sensor — 39
General information — 31
Oxygen sensor — 38
Relieving fuel pressure — 43
Removal and installation — 44
Speed sensor — 40
Thermo-time switch — 42
Throttle switch — 37
Throttle vacuum switch — 38
Troubleshooting — 33
Line pressure regulator — 54
Logic module — 119, 132

M

Manifold pressure sensor
Chrysler multi-port — 120
Chrysler single point — 133
Ford CFI — 96
General information — 15
GM digital — 87
Motronic (Bosch) fuel injection system
Air mass sensor — 36
Auxiliary air regulator — 42
Boost sensor — 40
Cold start valve — 41
Coolant temperature sensor — 39
Detonation sensor — 40
Electronic control unit — 35
Fuel injectors — 40
Fuel pressure regulator — 34
Fuel pump — 43
Fuel temperature sensor — 39

General information — 31
Oxygen sensor — 38
Relieving fuel pressure — 43
Removal and installation — 44
Speed sensor — 40
Thermo-time switch — 42
Throttle switch — 37
Throttle vacuum switch — 38
Troubleshooting — 33
Multi-port (Chrysler) fuel injection system
Coolant temperature sensor — 121
Detonation sensor — 122
Fuel injectors — 123
Fuel pressure regulator — 123
General description — 116
Idle air control — 123
Intake charge temperature sensor — 121
Logic module — 119
Manifold pressure sensor — 120
Oxygen sensor — 121
Power module — 120
Relieving fuel pressure — 124
Removal and installation — 124
Throttle body — 122
Throttle position sensor — 123
Troubleshooting — 117
Multi-port (Ford) fuel injection system
Air bypass valve — 109
Air flow meter — 108
Fuel injectors — 110
General description — 107
Oxygen sensor — 110
Relieving fuel pressure — 111
Removal and installation — 111
Throttle body — 109
Troubleshooting — 107
Multi-port (GM) fuel injection system
Air mass sensor — 77
Electronic control module — 77
Fuel pump — 77
Idle air control — 77
General information — 75
Relieving fuel pressure — 78
Removal and installation — 78
Troubleshooting — 76

O

Oxygen sensor
Chrysler multi-port — 121
Chrysler single point — 133
Ford CFI — 95
Ford multi-port — 110
General information — 13
GM multi-port — 79
K/KE-Jetronic — 55
L/LH-Jetronic — 38
Motronic — 38

Index

P

Power module — 120, 132

R

Relieving fuel pressure
Chrysler multi-port — 124
Chrysler single point — 136
Ford CFI — 98
Ford multi-port — 111
GM digital — 87
GM multi-port — 78
GM sequential — 89
GM-TBI — 66
K/KE-Jetronic — 59
L/LH-Jetronic — 43
Motronic — 43
Removal and installation
Chrysler multi-port — 124
Chrysler single point — 136
Ford CFI — 99
Ford multi-port — 124
GM digital — 87
GM multi-port — 78
GM sequential — 89
GM-TBI — 67
K/KE-Jetronic — 59
L/LH-Jetronic — 44
Motronic — 44

S

Sensors — 11
Sequential (GM) fuel injection system
Air mass sensor — 88
Electronic control module — 88
Fuel pump — 89
General information — 87
Idle air control — 89
Relieving fuel pressure — 89
Single point (Chrysler) fuel injection system
Automatic idle speed control — 135
Coolant temperature sensor — 134
Fuel injector — 135
Fuel pressure regulator — 134
General description — 129
Logic module — 132
Manifold pressure sensor — 133
Oxygen sensor — 133
Power module — 132
Relieving fuel pressure — 136
Removal and installation — 136
Throttle body — 134
Throttle body temperature sensor — 134

Throttle position switch — 134
Troubleshooting — 130
Speed sensor
General information — 15
L/LH-Jetronic — 40
Motronic — 40
System applications — 23

T

Thermo-time switch
General information — 20
K/KE-Jetronic — 56
L/LH-Jetronic — 42
Motronic — 42
Throttle body
Chrysler multi-port — 122
Chrysler single point — 134
Ford CFI — 97
Ford multi-port — 109
GM digital — 85
GM-TBI — 63
Throttle body (GM) fuel injection system
Electronic control module — 65
Fuel pressure regulator — 66
Fuel pump — 66
General information — 63
Idle air control valve — 65
Relieving fuel pressure — 66
Removal and installation — 67
Troubleshooting — 63
Throttle body injection — 10
Throttle body temperature sensor — 134
Throttle position sensor
Chrysler multi-port — 123
Chrysler single point — 134
Ford CFI — 95
General information — 12
GM digital — 86
GM multi-port — 79, 84
GM-TBI — 68, 73
Throttle switch — 37
Throttle vacuum switch — 38
Throttle valve — 54
Troubleshooting
Chrysler multi-port — 117
Chrysler single point — 130
Ford CFI — 91
Ford multi-port — 107
GM multi-port — 76
GM-TBI — 63
K/KE-Jetronic — 51
L/LH-Jetronic — 33
Motronic — 33
Types of fuel injection — 8

HAYNES AUTOMOTIVE MANUALS

NOTE: New manuals are added to this list on a periodic basis. If you do not see a listing for your vehicle, consult your local Haynes dealer for the latest product information.

ACURA
- *1776 **Integra & Legend** all models '86 thru '90

AMC
- **Jeep CJ** - see JEEP (412)
- 694 **Mid-size models**, Concord, Hornet, Gremlin & Spirit '70 thru '83
- 934 **(Renault) Alliance & Encore** all models '83 thru '87

AUDI
- 615 **4000** all models '80 thru '87
- 428 **5000** all models '77 thru '83
- 1117 **5000** all models '84 thru '88

AUSTIN
- **Healey Sprite** - see MG Midget Roadster (265)

BMW
- *2020 **3/5 Series** not including diesel or all-wheel drive models '82 thru '92
- 276 **320i** all 4 cyl models '75 thru '83
- 632 **528i & 530i** all models '75 thru '80
- 240 **1500 thru 2002** all models except Turbo '59 thru '77
- 348 **2500, 2800, 3.0 & Bavaria** all models '69 thru '76

BUICK
- **Century (front wheel drive)** - see GENERAL MOTORS (829)
- *1627 **Buick, Oldsmobile & Pontiac Full-size (Front wheel drive)** all models '85 thru '93
  Buick Electra, LeSabre and Park Avenue;
  Oldsmobile Delta 88 Royale, Ninety Eight and Regency; **Pontiac** Bonneville
- 1551 **Buick Oldsmobile & Pontiac Full-size (Rear wheel drive)**
  Buick Estate '70 thru '90, Electra '70 thru '84, LeSabre '70 thru '85, Limited '74 thru '79
  Oldsmobile Custom Cruiser '70 thru '90, Delta 88 '70 thru '85,Ninety-eight '70 thru '84
  Pontiac Bonneville '70 thru '81, Catalina '70 thru '81, Grandville '70 thru '75, Parisienne '83 thru '86
- 627 **Mid-size Regal & Century** all rear-drive models with V6, V8 and Turbo '74 thru '87
  Regal - see GENERAL MOTORS (1671)
  Skyhawk - see GENERAL MOTORS (766)
- 552 **Skylark** all X-car models '80 thru '85
  Skylark '86 on - see GENERAL MOTORS (1420)
  Somerset - see GENERAL MOTORS (1420)

CADILLAC
- *751 **Cadillac Rear Wheel Drive** all gasoline models '70 thru '92
  Cimarron - see GENERAL MOTORS (766)

CAPRI
- 296 **2000 MK I Coupe** all models '71 thru '75
  Mercury Capri - see FORD Mustang (654)

CHEVROLET
- *1477 **Astro & GMC Safari Mini-vans** '85 thru '93
- 554 **Camaro V8** all models '70 thru '81
- 866 **Camaro** all models '82 thru '92
  Cavalier - see GENERAL MOTORS (766)
  Celebrity - see GENERAL MOTORS (829)
- 625 **Chevelle, Malibu & El Camino** all V6 & V8 models '69 thru '87
- 449 **Chevette & Pontiac T1000** '76 thru '87
- 550 **Citation** all models '80 thru '85

- *1628 **Corsica/Beretta** all models '87 thru '92
- 274 **Corvette** all V8 models '68 thru '82
- *1336 **Corvette** all models '84 thru '91
- 1762 **Chevrolet Engine Overhaul Manual**
- 704 **Full-size Sedans** Caprice, Impala, Biscayne, Bel Air & Wagons '69 thru '90
  Lumina - see GENERAL MOTORS (1671)
  Lumina APV - see GENERAL MOTORS (2035)
- 319 **Luv Pick-up** all 2WD and 4WD '72 thru '82
- 626 **Monte Carlo** all models '70 thru '88
- 241 **Nova** all V8 models '69 thru '79
- *1642 **Nova and Geo Prizm** all front wheel drive models, '85 thru '92
- 420 **Pick-ups '67 thru '87** - Chevrolet & GMC, all V8 & in-line 6 cyl, 2WD & 4WD '67 thru '87; Suburbans, Blazers & Jimmys '67 thru '91
- *1664 **Pick-ups '88 thru '93** - Chevrolet & GMC, all full-size (C and K) models, '88 thru '93
- *831 **S-10 & GMC S-15 Pick-ups** all models '82 thru '92
- *1727 **Sprint & Geo Metro** '85 thru '91
- *345 **Vans - Chevrolet & GMC**, V8 & in-line 6 cylinder models '68 thru '92

CHRYSLER
- *2058 **Full-size Front-Wheel Drive** '88 thru '93
  K-Cars - see DODGE Aries (723)
  Laser - see DODGE Daytona (1140)
- *1337 **Chrysler & Plymouth Mid-size** front wheel drive '82 thru '93

DATSUN
- 402 **200SX** all models '77 thru '79
- 647 **200SX** all models '80 thru '83
- 228 **B - 210** all models '73 thru '78
- 525 **210** all models '78 thru '82
- 206 **240Z, 260Z & 280Z** Coupe '70 thru '78
- 563 **280ZX** Coupe & 2+2 '79 thru '83
  300ZX - see NISSAN (1137)
- 679 **310** all models '78 thru '82
- 123 **510 & PL521 Pick-up** '68 thru '73
- 430 **510** all models '78 thru '81
- 372 **610** all models '72 thru '76
- 277 **620 Series Pick-up** all models '73 thru '79
  720 Series Pick-up - see NISSAN (771)
- 376 **810/Maxima** all gasoline models, '77 thru '84
- 368 **F10** all models '76 thru '79
  Pulsar - see NISSAN (876)
  Sentra - see NISSAN (982)
  Stanza - see NISSAN (981)

DODGE
- 400 & 600 - see CHRYSLER Mid-size (1337)
- *723 **Aries & Plymouth Reliant** '81 thru '89
- *1231 **Caravan & Plymouth Voyager Mini-Vans** all models '84 thru '93
- 699 **Challenger & Plymouth Saporro** all models '78 thru '83
  Challenger '67-'76 - see DODGE Dart (234)
- 236 **Colt** all models '71 thru '77
- 610 **Colt & Plymouth Champ (front wheel drive)** all models '78 thru '87
- *1668 **Dakota Pick-ups** all models '87 thru '93
- 234 **Dart, Challenger/Plymouth Barracuda & Valiant** 6 cyl models '67 thru '76
- *1140 **Daytona & Chrysler Laser** '84 thru '89
- *545 **Omni & Plymouth Horizon** '78 thru '90
- *912 **Pick-ups** all full-size models '74 thru '91
- *556 **Ram 50/D50 Pick-ups & Raider and Plymouth Arrow Pick-ups** '79 thru '93
- *1726 **Shadow & Plymouth Sundance** '87 thru '93
- *1779 **Spirit & Plymouth Acclaim** '89 thru '92
- *349 **Vans - Dodge & Plymouth** V8 & 6 cyl models '71 thru '91

EAGLE
- **Talon** - see Mitsubishi Eclipse (2097)

FIAT
- 094 **124 Sport Coupe & Spider** '68 thru '78
- 273 **X1/9** all models '74 thru '80

FORD
- *1476 **Aerostar Mini-vans** all models '86 thru '92
- 788 **Bronco and Pick-ups** '73 thru '79
- *880 **Bronco and Pick-ups** '80 thru '91
- 268 **Courier Pick-up** all models '72 thru '82
- 1763 **Ford Engine Overhaul Manual**
- 789 **Escort/Mercury Lynx** all models '81 thru '90
- *2046 **Escort/Mercury Tracer** '91 thru '93
- *2021 **Explorer & Mazda Navajo** '91 thru '92
- 560 **Fairmont & Mercury Zephyr** '78 thru '83
- 334 **Fiesta** all models '77 thru '80
- 754 **Ford & Mercury Full-size,**
  Ford LTD & Mercury Marquis ('75 thru '82);
  Ford Custom 500,Country Squire, Crown Victoria & Mercury Colony Park ('75 thru '87);
  Ford LTD Crown Victoria & Mercury Gran Marquis ('83 thru '87)
- 359 **Granada & Mercury Monarch** all in-line, 6 cyl & V8 models '75 thru '80
- 773 **Ford & Mercury Mid-size,**
  Ford Thunderbird & Mercury Cougar ('75 thru '82);
  Ford LTD & Mercury Marquis ('83 thru '86);
  Ford Torino,Gran Torino, Elite, Ranchero pick-up, LTD II, Mercury Montego, Comet, XR-7 & Lincoln Versailles ('75 thru '86)
- *654 **Mustang & Mercury Capri** all models including Turbo.
  Mustang, '79 thru '92; Capri, '79 thru '86
- 357 **Mustang V8** all models '64-1/2 thru '73
- 231 **Mustang II** 4 cyl, V6 & V8 models '74 thru '78
- 649 **Pinto & Mercury Bobcat** '75 thru '80
- 1670 **Probe** all models '89 thru '92
- *1026 **Ranger/Bronco II** gasoline models '83 thru '93
- *1421 **Taurus & Mercury Sable** '86 thru '92
- *1418 **Tempo & Mercury Topaz** all gasoline models '84 thru '93
- 1338 **Thunderbird/Mercury Cougar** '83 thru '88
- *1725 **Thunderbird/Mercury Cougar** '89 and '90
- *344 **Vans** all V8 Econoline models '69 thru '91

GENERAL MOTORS
- *829 **Buick Century, Chevrolet Celebrity, Oldsmobile Cutlass Ciera & Pontiac 6000** all models '82 thru '93
- *766 **Buick Skyhawk, Cadillac Cimarron, Chevrolet Cavalier, Oldsmobile Firenza & Pontiac J-2000 & Sunbird** all models '82 thru '92
- 1420 **Buick Skylark & Somerset, Oldsmobile Calais & Pontiac Grand Am** all models '85 thru '91
- *1671 **Buick Regal, Chevrolet Lumina, Oldsmobile Cutlass Supreme & Pontiac Grand Prix** all front wheel drive models '88 thru '90
- *2035 **Chevrolet Lumina APV, Oldsmobile Silhouette & Pontiac Trans Sport** all models '90 thru '92

GEO
- **Metro** - see CHEVROLET Sprint (1727)
- **Prizm** - see CHEVROLET Nova (1642)
- *2039 **Storm** all models '90 thru '92
- **Tracker** - see SUZUKI Samurai (1626)

GMC
- **Safari** - see CHEVROLET ASTRO (1477)
- **Vans & Pick-ups** - see CHEVROLET (420, 831, 345, 1664)

(Continued on other side)

---

* *Listings shown with an asterisk (*) indicate model coverage as of this printing. These titles will be periodically updated to include later model years - consult your Haynes dealer for more information.*

Haynes North America, Inc., 861 Lawrence Drive, Newbury Park, CA 91320 • (805) 498-6703

Cape May 609 884 2736
Blvd Obsen

HAYNES AUTOMOTIVE MANUALS

NOTE: New manuals are added to this list on a periodic basis. If you do not see a listing for your vehicle, consult your local Haynes dealer for the latest product information.

HONDA
- 351 Accord CVCC all models '76 thru '83
- 1221 Accord all models '84 thru '89
- 2067 Accord all models '90 thru '93
- 160 Civic 1200 all models '73 thru '79
- 633 Civic 1300 & 1500 CVCC all models '80 thru '83
- 297 Civic 1500 CVCC all models '75 thru '79
- 1227 Civic all models '84 thru '91
- *601 Prelude CVCC all models '79 thru '89

HYUNDAI
- *1552 Excel all models '86 thru '93

ISUZU
- *1641 Trooper & Pick-up, all gasoline models Pick-up, '81 thru '93; Trooper, '84 thru '91

JAGUAR
- *242 XJ6 all 6 cyl models '68 thru '86
- *478 XJ12 & XJS all 12 cyl models '72 thru '85

JEEP
- *1553 Cherokee, Comanche & Wagoneer Limited all models '84 thru '93
- 412 CJ all models '49 thru '86
- *1777 Wrangler all models '87 thru '92

LADA
- *413 1200, 1300. 1500 & 1600 all models including Riva '74 thru '91

MAZDA
- 648 626 Sedan & Coupe (rear wheel drive) all models '79 thru '82
- *1082 626 & MX-6 (front wheel drive) all models '83 thru '91
- 267 B Series Pick-ups '72 thru '93
- 370 GLC Hatchback (rear wheel drive) all models '77 thru '83
- 757 GLC (front wheel drive) '81 thru '85
- *2047 MPV all models '89 thru '93
- 460 RX-7 all models '79 thru '85
- *1419 RX-7 all models '86 thru '91

MERCEDES-BENZ
- *1643 190 Series all four-cylinder gasoline models, '84 thru '88
- 346 230, 250 & 280 Sedan, Coupe & Roadster all 6 cyl sohc models '68 thru '72
- 983 280 123 Series gasoline models '77 thru '81
- 698 350 & 450 Sedan, Coupe & Roadster all models '71 thru '80
- 697 Diesel 123 Series 200D, 220D, 240D, 240TD, 300D, 300CD, 300TD, 4- & 5-cyl incl. Turbo '76 thru '85

MERCURY
See FORD Listing

MG
- 111 MGB Roadster & GT Coupe all models '62 thru '80
- 265 MG Midget & Austin Healey Sprite Roadster '58 thru '80

MITSUBISHI
- *1669 Cordia, Tredia, Galant, Precis & Mirage '83 thru '93
- *2022 Pick-up & Montero '83 thru '93
- *2097 Eclipse, Eagle Talon & Plymouth Laser '90 thru '94

MORRIS
- 074 (Austin) Marina 1.8 all models '71 thru '78
- 024 Minor 1000 sedan & wagon '56 thru '71

NISSAN
- 1137 300ZX all models including Turbo '84 thru '89
- *1341 Maxima all models '85 thru '91
- *771 Pick-ups/Pathfinder gas models '80 thru '93
- 876 Pulsar all models '83 thru '86
- *982 Sentra all models '82 thru '90
- *981 Stanza all models '82 thru '90

OLDSMOBILE
- Bravada - see CHEVROLET S-10 (831)
- Calais - see GENERAL MOTORS (1420)
- Custom Cruiser - see BUICK Full-size RWD (1551)
- *658 Cutlass all standard gasoline V6 & V8 models '74 thru '88
- Cutlass Ciera - see GENERAL MOTORS (829)
- Cutlass Supreme - see GM (1671)
- Delta 88 - see BUICK Full-size RWD (1551)
- Delta 88 Brougham - see BUICK Full-size FWD (1627), RWD (1551)
- Delta 88 Royale - see BUICK Full-size RWD (1551)
- Firenza - see GENERAL MOTORS (766)
- Ninety-eight Regency - see BUICK Full-size RWD (1551), FWD (1627)
- Ninety-eight Regency Brougham - see BUICK Full-size RWD (1551)
- Omega - see PONTIAC Phoenix (551)
- Silhouette - see GENERAL MOTORS (2035)

PEUGEOT
- 663 504 all diesel models '74 thru '83

PLYMOUTH
- Laser - see MITSUBISHI Eclipse (2097)
- For other PLYMOUTH titles, see DODGE listing.

PONTIAC
- T1000 - see CHEVROLET Chevette (449)
- J-2000 - see GENERAL MOTORS (766)
- 6000 - see GENERAL MOTORS (829)
- Bonneville - see Buick Full-size FWD (1627), RWD (1551)
- Bonneville Brougham - see Buick Full-size (1551)
- Catalina - see Buick Full-size (1551)
- 1232 Fiero all models '84 thru '88
- 555 Firebird V8 models except Turbo '70 thru '81
- 867 Firebird all models '82 thru '92
- Full-size Rear Wheel Drive - see BUICK Oldsmobile, Pontiac Full-size RWD (1551)
- Full-size Front Wheel Drive - see BUICK Oldsmobile, Pontiac Full-size FWD (1627)
- Grand Am - see GENERAL MOTORS (1420)
- Grand Prix - see GENERAL MOTORS (1671)
- Grandville - see BUICK Full-size (1551)
- Parisienne - see BUICK Full-size (1551)
- 551 Phoenix & Oldsmobile Omega all X-car models '80 thru '84
- Sunbird - see GENERAL MOTORS (766)
- Trans Sport - see GENERAL MOTORS (2035)

PORSCHE
- *264 911 all Coupe & Targa models except Turbo & Carrera 4 '65 thru '89
- 239 914 all 4 cyl models '69 thru '76
- 397 924 all models including Turbo '76 thru '82
- *1027 944 all models including Turbo '83 thru '89

RENAULT
- 141 5 Le Car all models '76 thru '83
- 079 8 & 10 58.4 cu in engines '62 thru '72
- 097 12 Saloon & Estate 1289 cc engine '70 thru '80
- 768 15 & 17 all models '73 thru '79
- 081 16 89.7 cu in & 95.5 cu in engines '65 thru '72
- Alliance & Encore - see AMC (934)

SAAB
- 247 99 all models including Turbo '69 thru '80
- *980 900 all models including Turbo '79 thru '88

SUBARU
- 237 1100, 1300, 1400 & 1600 '71 thru '79
- *681 1600 & 1800 2WD & 4WD '80 thru '89

SUZUKI
- *1626 Samurai/Sidekick and Geo Tracker all models '86 thru '93

TOYOTA
- 1023 Camry all models '83 thru '91
- 150 Carina Sedan all models '71 thru '74
- 935 Celica Rear Wheel Drive '71 thru '85
- *2038 Celica Front Wheel Drive '86 thru '92
- 1139 Celica Supra all models '79 thru '92
- 361 Corolla all models '75 thru '79
- 961 Corolla all rear wheel drive models '80 thru '87
- *1025 Corolla all front wheel drive models '84 thru '92
- 636 Corolla Tercel all models '80 thru '82
- 360 Corona all models '74 thru '82
- 532 Cressida all models '78 thru '82
- 313 Land Cruiser all models '68 thru '82
- 200 MK II all 6 cyl models '72 thru '76
- *1339 MR2 all models '85 thru '87
- 304 Pick-up all models '69 thru '78
- *656 Pick-up all models '79 thru '92
- *2048 Previa all models '91 thru '93

TRIUMPH
- 112 GT6 & Vitesse all models '62 thru '74
- 113 Spitfire all models '62 thru '81
- 322 TR7 all models '75 thru '81

VW
- 159 Beetle & Karmann Ghia all models '54 thru '79
- 238 Dasher all gasoline models '74 thru '81
- *884 Rabbit, Jetta, Scirocco, & Pick-up gas models '74 thru '91 & Convertible '80 thru '92
- 451 Rabbit, Jetta & Pick-up all diesel models '77 thru '84
- 082 Transporter 1600 all models '68 thru '79
- 226 Transporter 1700, 1800 & 2000 all models '72 thru '79
- 084 Type 3 1500 & 1600 all models '63 thru '73
- 1029 Vanagon all air-cooled models '80 thru '83

VOLVO
- 203 120, 130 Series & 1800 Sports '61 thru '73
- 129 140 Series all models '66 thru '74
- *270 240 Series all models '74 thru '90
- 400 260 Series all models '75 thru '82
- *1550 740 & 760 Series all models '82 thru '88

SPECIAL MANUALS
- 1479 Automotive Body Repair & Painting Manual
- 1654 Automotive Electrical Manual
- 1667 Automotive Emissions Control Manual
- 1480 Automotive Heating & Air Conditioning Manual
- 1762 Chevrolet Engine Overhaul Manual
- 1736 GM and Ford Diesel Engine Repair Manual
- 1763 Ford Engine Overhaul Manual
- 482 Fuel Injection Manual
- 2069 Holley Carburetor Manual
- 1666 Small Engine Repair Manual
- 299 SU Carburetors thru '88
- 393 Weber Carburetors thru '79
- 300 Zenith/Stromberg CD Carburetors thru '76

* Listings shown with an asterisk (*) indicate model coverage as of this printing. These titles will be periodically updated to include later model years - consult your Haynes dealer for more information.

Over 100 Haynes motorcycle manuals also available

5-94

Haynes North America, Inc., 861 Lawrence Drive, Newbury Park, CA 91320 • (805) 498-6703